Human Thermal Comfort

Human Thermal Comfort

Ken Parsons

CRC Press
Taylor & Francis Group
Boca Raton London New York

CRC Press is an imprint of the
Taylor & Francis Group, an **informa** business

CRC Press
Taylor & Francis Group
6000 Broken Sound Parkway NW, Suite 300
Boca Raton, FL 33487-2742

© 2020 by Taylor & Francis Group, LLC
CRC Press is an imprint of Taylor & Francis Group, an Informa business

No claim to original U.S. Government works

Printed on acid-free paper

International Standard Book Number-13: 978-0-367-26193-1 (Hardback)

Library of Congress Cataloging-in-Publication Data

Names: Parsons, K. C. (Kenneth C.), 1953- author.
Title: Human thermal comfort / by Ken Parsons.
Description: Boca Raton, FL: CRC Press/Taylor & Francis Group 2020. |
Includes bibliographical references and index.
Identifiers: LCCN 2019040885 (print) | LCCN 2019040886 (ebook) |
ISBN 9780367261931 (hardback; acid-free paper) |
ISBN 9780429294983 (ebook)
Subjects: LCSH: Temperature—Physiological effect—Mathematical
models. | Environmental engineering.
Classification: LCC QP82.2.T4 P379 2020 (print) | LCC QP82.2.T4 (ebook) |
DDC 612/.01426—dc23
LC record available at https://lccn.loc.gov/2019040885
LC ebook record available at https://lccn.loc.gov/2019040886

Visit the Taylor & Francis Web site at
http://www.taylorandfrancis.com

and the CRC Press Web site at
http://www.crcpress.com

To Jane, Ben, Anna, Hannah, Richard, Sam,
Mina, Nancy, Thomas and Edward

Contents

Preface

As we progress through the 21st century, we know a great deal about conditions that create thermal comfort and the causes of discomfort. For practical application, there have been so many studies, conferences, debates and there is so much data and thousands of academic papers and reports into thermal comfort that it seems like we have enough. However, as almost any environmental survey that considers human comfort and well-being will show, discomfort and dissatisfaction due to being too hot or too cold remains a main source of complaint among people worldwide.

Thermal comfort has always been a goal for people (one of the reasons why stone-age men and women lived in caves, wore clothing or lit fires). Only with the industrial revolution and the development of the scientific method in the 19th century have formal studies and integrated and shared knowledge emerged. This development can be categorized into four overlapping phases with a fifth and maybe more yet to come.

The first phase of 'formal' investigations into thermal comfort progressed with developments in heating, ventilation and later air conditioning. It investigated thermal comfort using the mainly subjective responses of people, in the laboratory and in the 'field', to a variety of thermal conditions. This was the pre-computer age. The second phase arrived at around the 1960s and was based upon the heat transfer between a person and the thermal environment. Body heat equations were developed and linked with the results of thermal comfort research using people, to provide universal methods that predicted thermal comfort responses in all thermal environments. That is, not restricted by the limited number of conditions that could be investigated using human subjects. At first, these methods were considered too complex, however, 'rescued' by the coming of the computer, they launched us into a new age still dominant today.

The third phase of thermal comfort investigation is computer simulation where computer models of human thermoregulation, heat transfer and the thermal properties of the human body were integrated into computer models for use in computer-aided environmental design and assessment. To a large extent, these models superseded the use of simple heat transfer models, however they have only partly been accepted by standards and regulations. Physical models could also be included here where body-shaped heated manikins are physical simulations used to evaluate thermal environments as well as for measuring instruments to determine heat transfer particularly through clothing. A natural progression of this work is into moving, sweating, heated manikins and robotics. The fourth phase of development is the addition of human behaviour as essential to an understanding of thermal comfort leading to what are called adaptive models. We can speculate about a fifth phase and beyond. The very large number of studies into thermal comfort lends itself to what I called database modelling (Parsons and Bishop, 1991) and is related to meta-analysis and 'big data'. It may be in the future that databases could become so large and search engines so efficient that with standardised collection methods, predictions can be totally empirical and hence from real people. Any evaluation or prediction involving thermal comfort could be obtained by interrogating previously determined data identical (matched) to the environment and people of interest. Before that however, the fifth phase must integrate all methods involving biophysics, heat transfer, anatomy and physiology and human behaviour to predict thermal comfort in all environments for all people from female children with disabilities in the USA to male football players in Japan and all between. This focus book presents up-to-date current knowledge of thermal comfort and contributes to laying foundations for the future. We start with phase two and move forward.

Ole Fanger, a young engineer from Denmark, in his seminal work 'Thermal Comfort' (1970), noted that despite its fundamental importance to human existence and to the heating, ventilation and air conditioning industry "…existing knowledge of thermal comfort was quite inadequate and unsuitable for practical applications." Fanger (1970).

The methods he presented were a paradigm shift in how to specify thermal comfort conditions and have gone on to be accepted worldwide and to form the basis of national, regional and international standards. This is why this focus book begins with a chapter that introduces the topic of thermal comfort but moves on immediately to describe conditions for comfort captured by Fanger (1970) in a comfort equation. It then moves to the Predicted Mean Vote (PMV) and the associated Predicted Percentage of Dissatisfied (PPD) thermal comfort indices first proposed by Professor Fanger (1970) in his book and that still form the basis of current methods. Chapter 4 then describes the work of Professor Pharo Gagge and his colleagues at the J B Pierce Foundation Laboratory in Yale, that was influential in the USA. The New Effective Temperature (ET*) and the Standard Effective Temperature (SET) were similar to the PMV/PPD indices in derivation but used a thermal model and equivalent environment approach to provide the indices. Thermal discomfort can be caused by disturbance to parts of the body as well as overall. Chapter 5 considers local thermal discomfort and describes sensations when people are exposed to thermal conditions specific to parts of the body such as the hands, head, neck and feet. Chapter 5 considers discomfort caused by draughts, asymmetric radiation, vertical temperature gradients and contact with surfaces.

This focus book then moves to the next stage of thermal comfort research. The ingredients were available through ideas and research conducted since 1960 and have been developed, particularly throughout the 1990s into the next paradigm shift in this area. Chapter 6 considers the topic of adaptive thermal comfort. The paradigm shift from the work of Fanger (1970) to the present day has been the introduction of methods that acknowledge that people are not static in their reception of thermal environments and that they adapt or behave to maintain and achieve thermal comfort. Chapter 6 builds on the earlier chapters to integrate these ideas into a new approach. Chapter 6 then moves on to a shift in models of human thermoregulation from an automatic system, based upon physiology, that underpins existing methods to embracing research into behavioural thermoregulation and adaptive methods to provide a model of human thermoregulation. This new model integrates both parts of the system into a realistic representation of how people respond to achieve thermal comfort when they begin to experience thermal discomfort and dissatisfaction.

Much of research into thermal comfort is into relatively steady-state conditions, however, there are specific environments such as those in vehicles and those outdoors where this may not exist. Chapter 7 considers thermal comfort in special environments, and Chapter 8 considers thermal comfort for special populations. That is for those outside of the commonly researched population to provide a consideration of diversity for different peoples and cultures across the world and people with disabilities.

Chapter 9 considers human performance in thermal environments as they move from optimum performance when comfortable to causing a drop in performance when people become too hot or too cold. This is described using the HSDC approach (Health and Safety, Distraction and Capacity). Chapter 10 describes international standards for thermal comfort assessment and also how to construct a spreadsheet 'macro' computer model of the internationally standardised PMV and PPD indices. Chapter 11 demonstrates how the new model of thermal comfort can be used in practical application as well as presenting an example of a method for assessing an existing environment for thermal comfort.

Ken Parsons
August 2019
Loughborough, United Kingdom

Author

Ken Parsons is Emeritus Professor of Environmental Ergonomics at Loughborough University. He has spent over 30 years conducting laboratory and field research into human thermal comfort. He was born on January 20, 1953, in northeast England in a coastal village called Seaton Sluice. He graduated from Loughborough University in ergonomics in 1974, obtained a postgraduate certificate in education in mathematics with a distinction from Hughes Hall, Cambridge University in 1975 and was awarded a PhD in human response to vibration in 1980, from the Institute of Sound and Vibration Research, Southampton University. He founded the Human Thermal Environments Laboratory at Loughborough in 1981 and was awarded a certificate in management from the Open University in 1993. Ken became head of the Department of Human Sciences in 1996 covering research and teaching in ergonomics, psychology and human biology. He was Dean of Science from 2003 to 2009 and Pro-Vice-Chancellor for research from 2009 to 2012. He was chair of the United Kingdom Deans of Science from 2008 to 2010.

In 1992, he received the Ralph G Nevins Award from the American Society of Heating, Refrigerating and Air-Conditioning Engineers (ASHRAE) for 'significant accomplishments in the study of bioenvironmental engineering and its impact on human comfort and health'. The Human Thermal Environments Laboratory was awarded the President's Medal of the Ergonomics Society in 2001. He is one of the co-authors of the British Occupational Hygiene Society publication on thermal environments and has contributed to the Chartered Institute of Building Services Engineers publications on thermal comfort as well as to the *ASHRAE Handbook: Fundamentals*.

He has been a fellow of the Institute of Ergonomics and Human Factors, the International Ergonomics Association and the Royal Society of Medicine. He was a registered European Ergonomist and an elected member to the council of the Ergonomics Society. He has been a scientific advisor to the Defence Evaluation Research Agency and the Defence Clothing and Textile Agency and a member of the Defence Scientific Advisory Committee. He has been both secretary and chair of the thermal factors committee of the International Commission on Occupational Health (ICOH), chair of the Centre National de la Recherche Scientifique (CNRS) advisory committee to the Laboratoire de Physiologie et Psychologie Environmentales in Strasbourg, France, and is a life member of the Indian Ergonomics Society. He was a visiting professor to Chalmers University in Sweden and is a member of the committee of the International Conference on Environmental Ergonomics. He was an advisor to the World Health Organization on heatwaves and a visiting professor to Chongqing University in China, where he was leading academic to the National Centre for International Research of Low Carbon and Green Buildings. He was scientific editor and co-editor in chief of the journal *Applied Ergonomics* for 33 years and is on the editorial boards of the journals *Industrial Health, Annals of Occupational Hygiene* and *Physiological Anthropology*.

He is co-founder of the United Kingdom Indoor Environments Group and a founding member of the UK Clothing Science Group, the European Society for Protective Clothing, the Network for Comfort and Energy Use in Buildings and the thermal factors scientific committee of the ICOH. He was chair of ISO TC 159 SC5 'Ergonomics of the Physical Environment' for over 20 years and is convenor to the ISO working group on integrated environments, chair of the British Standards Institution committee on the ergonomics of the physical environment and convenor of CEN TC 122 WG11, which is the European standards committee concerned with the ergonomics of the physical environment.

Human Thermal Comfort

1

SEVEN FACTORS FOR THERMAL COMFORT

There are seven factors that must be taken into account when considering human thermal comfort. These are the air temperature, radiant temperature, humidity and air velocity to which a person is exposed, the metabolic heat which is produced by their activity, the clothing they wear and the adaptive opportunities afforded by the environment they occupy including their capabilities of taking advantage of them.

There are additional factors for some environments that require special consideration. For example in hypo- or hyper-baric environments such as up mountains, in aircraft or under water where pressure will also be important and in space where gravity will be absent or on other planets where it will vary from that on earth. The seven factors therefore provide a minimum set when designing environments for thermal comfort. All of them must be taken into account and specification of thermal comfort conditions cannot be provided in terms of one or a sub-set of the seven factors. The interaction of the first six factors above is well researched in terms of thermal comfort. Designing for adaptive opportunities is in its infancy in terms of environmental design and assessment for thermal comfort and provides major opportunities for innovative ideas to achieve thermal comfort as well as for stimulating, delightful, pleasant and energy efficient thermal environments.

DIVERSE POPULATIONS AND ENVIRONMENTS

I gave a presentation on thermal comfort at an international conference, organised by the European Union, in Bruges, Belgium, in the late 1990s. The speaker before me had considered falling population levels in industrialized countries. From Japan, Korea and China in the east through Europe, including traditional catholic countries, to the USA, couples had borne, on average, less than two children. The conclusion that was identified for Europe, was that to provide for a sustainable work force and general population, migration of millions of people from Africa, the closest continent with population growth, was required. This has now happened and as a consequence of this migration, electronic communication including social media, and many other factors and population movements, globalisation has occurred and all of us live in, and benefit from, rich and diverse populations and cultures. Within and across these cultures there is diversity. As well as people of different size and age, there are people who are ill, people with disabilities, a range of genders, people who take drugs and more. A challenge for the 21st century is to design stimulating and pleasant indoor and outdoor environments in which all people from all parts of the globe, can live and that includes an understanding and provision of thermal comfort, a basic human requirement.

In my presentation I had considered international standards for thermal comfort across the world for future generations. After I had delivered my presentation, I was asked a question from the audience "What is the maximum temperature that should be allowed in offices?" This was an important question and involved, particularly economic, consequences ranging from when workers should stop working in hot weather to energy requirements for cooling in buildings to general issues of how to prepare for global warming. I did not hesitate to answer 33°C (91.4°F).

My immediate answer was a surprise to the chair and the audience as I had been clear in my presentation that thermal comfort for all people depends upon six variables (the discerning reader will have observed that I have now added a seventh factor). That is, not only air temperature, but also radiant temperature, air movement and humidity, which constitute the environmental variables, the integrated effects of which will determine thermal comfort at levels depending upon the activity level of the person and the clothing that they are wearing.

The reason I was prepared to give an answer as a single temperature was that most office workers do not sit in the sunshine without the ability to move around or adjust curtains, shutters or blinds and they generally do not sit or stand next to hot or cold machines or surfaces. Radiant temperatures will therefore be similar to air temperatures so a single temperature will represent both. Humidity in offices is not usually sufficient to significantly influence thermal comfort and air movement is not usually high, although opening windows and using fans are often options and are usually welcome as temperatures approach 33°C. Office work usually involves sedentary to light activity and clothing is usually light to a maximum thermal insulation of that of a business suit or standard uniform, and can often be reduced where there is no strict uniform code. So that is why a single temperature can be specified as a maximum value for offices. But why 33°C? (To be clear, this is not proposed as a temperature for comfort but a temperature that can be considered sufficiently uncomfortable to be unacceptable).

It is the human disposition that people respond to maintain an internal body temperature of around 37°C. This is part of what is termed homeostasis, which is an attempt to maintain relatively constant and optimum conditions (e.g. temperature) inside of the body for the internal organs and body functions to perform effectively. From extensive research across the world, it has been found that for comfort a person performing sedentary to light work, such as office work, will have an average skin temperature across the body of around 33°C. It has also been determined that for that level of activity, the body will produce between around 120 and 150 W (Joules per second) of heat. To maintain a constant internal body temperature, that heat must be lost to the environment otherwise the internal body temperature will rise. Details will be provided in later chapters of this book, however to cut a long story short, at an environmental temperature of 33°C and above, the heat would be lost by sweating. If the air temperature is at the required skin temperature then no heat can be lost by convection without a rise in skin temperature, and even then, heat must be lost mainly by the evaporation of sweat. It is reasonable to assume that, in office work, sweating becomes an unacceptable condition in terms of level of discomfort and stickiness in clothing as well as possible unacceptable stress and disagreeable appearance of face and clothing. It will also cause distraction and hence a possible reduction in time on task and general performance and productivity. It is probably reasonable to say that all people will not be comfortable at 33°C but no one will die of heat stress, and although other factors may come into play, 33°C is a good starting point for specifying maximum temperatures in offices worldwide.

For balance, we should consider minimum temperatures for offices and maybe homes. In this case, the ability to increase clothing levels is important, but using a similar rationale to that described above for maximum temperatures, around 18°C (64.4°F) seems to be a lower air temperature boundary for sedentary activity, as it is unacceptable to wear too much clothing insulation in offices and increased activity levels cease to be described as office work. Below 18°C, skin temperatures fall in an attempt by the body to reduce heat loss by convection and hence preserve metabolic heat to maintain internal body temperature. There is particular discomfort in the extremities of the body such as fingers, toes, ears and nose. That is because blood is withdrawn from the extremities by the body to preserve heat and the large surface area to mass (volume) ratio of the fingers and other extremities, promotes heat loss and reduces skin temperatures to uncomfortable levels. This is particularly important for people with thin fingers and explains why females are often more uncomfortable than males in cool conditions. Below 18°C, air temperature for sedentary office work would therefore start to become unacceptably uncomfortable, cause distraction and reduce performance at tasks involving fine dexterity as fingers will become stiff and cold.

A COMFORTABLE ENVIRONMENT

From the above discussion, it can be concluded that comfortable environments in offices or environments with light activity and light to moderate levels of clothing will not occur below temperatures of 18°C and above temperatures of 33°C. It has also been found that for office environments, air temperatures of around 23°C–24°C (73.4°F–75.2°F) would provide thermal comfort for most people. We must remember however, that these values have been determined for environments where assumptions have been made regarding general office environments. Fanger (1970) in his defining work *Thermal Comfort* notes that the most important variables that influence thermal comfort are activity level, thermal resistance of clothing, air temperature, mean radiant temperature, relative air velocity and water vapour pressure in ambient air. He concludes that it is the combined thermal effects of the variables that is important and that comfort can be achieved by many different combinations of the six variables. If activity levels and clothing can be reduced and air velocity levels increased, then comfort temperatures may move towards 33°C. If activity levels and clothing insulation can be increased and thermal radiation levels are high, then comfortable air temperatures can be well below 18°C. It is reasonable to assume that in offices, no one will die at (due to) air temperatures of 18°C but many will be uncomfortable and find it unacceptable. As an aside, it is worth noting in passing that cool air provides a feeling of freshness and usually provides better perceived air quality than that of warmer air. But that is not thermal comfort.

Comfort can be regarded as a desirable human condition and a comfortable environment one where people would require no change. In thermal terms, a person in thermal comfort would desire to be no hotter and no cooler.

There are many words, in many languages that can express this state as it is a universal human condition. It can be said that a person in thermal comfort is satisfied (or not dissatisfied) leading to the generally accepted definition of thermal comfort as "that condition of mind which expresses satisfaction with the thermal environment" (ASHRAE, 1966; Fanger, 1970).

This robust definition emphasises the psychological nature of thermal comfort. It has served the subject well over many years. More recent considerations of thermal comfort have however raised useful points particularly from an international perspective. The term 'condition of mind' is clear but how is it measured? Subjective responses are implied but maybe behavioural responses are more valid. No matter what a person says, if they move away from thermal conditions were they really comfortable? The expression of satisfaction may also have interpretation. The condition of mind which expresses satisfaction may be from a person who has a disposition not to complain; does not expect of even have experienced any 'better'; wishes to please the environmental designer, manager or the person conducting the survey or wishes to avoid the consequences of expressing dissatisfaction. The definition provided above is practical and probably valid in most contexts. In modern applications, however, we should be aware of the interpretations especially when suggestions are made that different populations and cultures require significantly different environments for thermal comfort.

PSYCHO-PHYSIOLOGICAL THERMOREGULATION AND COMFORT

Psycho-physiological thermoregulation is a rather long term, first formally proposed by Parsons (2019) but related to psychophysiological control and adaptive thermal comfort (e.g. Auliciems, 1981). It recognises that modern considerations of thermal comfort integrate both the continuous psychological or behavioural responses of people to thermal conditions as well as the continuous automatic or physiological responses of the body. Thermal conditions are the integration, at any point in time, of six variables that continuously change with time. These are the air temperature, radiant temperature, air velocity, humidity, clothing insulation and metabolic heat produced by activity.

The physiological system of thermoregulation detects the thermal state of the body, mainly at the skin (separate hot and cold sensors) but it also monitors internal conditions within the body, and this information is continuously transmitted to the brain (mainly the hypothalamus). The hypothalamus has two regions: the anterior hypothalamus responds when the body is too hot and changes the state of the skin by increasing blood supply to the skin surface (vasodilation) and if still required, by sweating. The posterior hypothalamus responds when the body is too cold and withdraws blood from the skin, especially the limbs, hands and feet (vasoconstriction). It also promotes tensing of the muscles in non-shivering thermogenesis and shivering that is an oscillatory contraction of muscle fibres (around 12 Hz starting in the neck) producing heat but not external work. It can initially be voluntary but in more severe conditions is involuntary. These effects can also be controlled 'locally' influencing the part of the skin surface affected and also contributing to the whole.

Blood absorbs the metabolic heat production due to activity and transports it around the body to the skin, which is the main surface for heat exchange. When a person is too hot, blood flows to the skin surface, raising its temperature and hence increasing potential loss of heat by convection from the body to the environment. Sweating wets the surface of the skin and increases greatly the potential for heat loss by evaporation. Withdrawal of the blood from the skin when the body is too cold reduces skin temperature and decreases the potential for heat loss by convection. Non-shivering thermogenesis and shivering increase metabolic heat. All physiological responses are an attempt to maintain internal body temperature at around 37°C. The magnitude of response is related to the thermal condition of the body. It is reasonable to assume that the strength of the response is related to a combination of internal body and skin temperatures, involving both temperature and rate of change of temperature, and the difference between set or desired temperatures and actual temperatures.

Investigations into the relationship between the physiological condition of the body and thermal comfort have concluded that if the body is in heat balance (maintaining internal body temperature at desired levels), then comfort and discomfort will relate to skin temperatures, leading to discomfort when skin temperatures fall and the body is cold, and sweat rates (or how wet the skin is) leading to stickiness and hence discomfort when the body is too hot. If skin temperatures and sweat rates are at 'optimum' levels then the body is in comfort.

The system of physiological thermoregulation described above provides the basis of methods used in the late 20th century, for determining conditions that create thermal comfort and predicting the degree of discomfort and dissatisfaction. More modern considerations include mostly conscious psychological factors in combination with the continuous and automatic unconscious physiological system described above. In brief, the body maintains homeostasis by influencing the environment it is exposed to by adjusting behaviour. If a person is uncomfortable and the opportunity is available, then, if disposed to do so, a person will adopt behaviours to preserve or produce heat or reduce heat loss if too cold; or if too hot, behave in a way to lose heat, reduce heat gain from the environment or reduce heat production. The obvious behaviours are to move away from an uncomfortable environment to a comfortable one, remove clothing when too hot, add clothing when too cold and change activity level, but there are many more. Behaviours adopted are often related to traditions and cultures of the people.

THERMAL COMFORT INDICES

A thermal comfort index is a single number that indicates a level of thermal comfort or discomfort. It is calculated from factors that influence thermal comfort, and for an ideal index, its value will change as thermal comfort changes. The factors that influence thermal comfort and are hence involved in the calculation of any comprehensive thermal comfort index are air temperature, radiant temperature, air velocity, humidity, clothing insulation and activity level. The index value is often taken as a temperature of a standard environment that provides equivalent thermal comfort to an actual environment. It can also be a prediction of a subjective state such as warm or cold and takes an appropriate value on a subjective scale (e.g. +2, warm; −1 slightly cool, etc.). A thermal comfort index has the practical utility of providing a single number to specify an environment for thermal comfort rather than needing to specify values of all six parameters each time a specification for a room, building or other space that is to be occupied by people is required.

There have been many attempts at determining the definitive thermal index that is valid, sensitive, reliable, can be used in practical applications and more. For a detailed description, the reader is referred to McIntyre (1980), Jendritzky et al. (2012) and Parsons (2014). There are essentially three types of index. Empirical indices are derived from using people to judge environments for thermal comfort, the index is then derived from the database of responses; rational indices use a calculation of heat transfer between the body and the environment and relate the predicted thermoregulatory responses (e.g. skin temperature, sweating, heat load, etc.) to empirically derived relationships between physiological states and thermal comfort. Direct indices are measuring instruments that respond to the variables that influence thermal comfort, and their values (usually temperature) are used as an index of thermal comfort (e.g. a black globe thermometer).

One of the first attempts to develop a thermal comfort index came from the American Society of Heating and Ventilating Engineers (ASHVE) who set up a laboratory in Pittsburgh in 1919 and studied the relative importance of air temperature and humidity on thermal comfort. This was eventually extended to include air velocity and thermal radiation to provide charts and nomograms of the relative importance to thermal warmth of each of the variables. The mechanism for deciding upon equivalence was for people to walk between two climatically controlled rooms set to different conditions. The people had to decide which room was warmer, and by conducting many trials, equivalence was found. The index was called the Effective Temperature (ET) and is the temperature of a room with still saturated air, which would give the same feeling of warmth as in the actual conditions under consideration (Houghton and Yagloglou, 1923). Later considerations included radiation (Vernon and Warner, 1932).

There have been a number of attempts internationally to derive thermal comfort indices (Resultant Temperature, Equivalent Temperature, Kata Cooling Power and more). The ET index was the most influential internationally. However, the recognition of its limitations, mainly due to the immediate responses of subjects rather than after their exposure for a prolonged period, more normal for rooms, but also the cumbersome nature of using charts, led to the thermal chambers moving from Pittsburgh to Kansas for more research. During the 1960s and early 1970s, extensive experiments were conducted with human subjects (e.g. Rohles and Nevins, 1971). Over this period, Ole Fanger, a young researcher from Denmark arrived in Kansas and inspired by the work at Kansas State University, as well as that of Pharo Gagge at the J B Pierce Laboratories, Yale and others, he changed everything.

It is essential to have an understanding of the work of P O Fanger if we wish to move forward to develop thermal comfort knowledge and methods of application into the future. This is described in detail in Chapter 2.

THERMAL COMFORT STANDARDS

Before moving on, we should note that there are many thermal comfort standards and regulations used throughout the world. Most are mainly to provide the specification for buildings although can be used more generally. Two standards in particular are internationally accepted. The standard ASHRAE[1] 55 is continuously reviewed and revised and is the accepted standard for specifying

---

[1] American Society of Heating, Refrigerating and Air Conditioning Engineers.

thermal comfort conditions. It provides thermal comfort conditions, usually presented on a psychrometric chart, along with other criteria, guidance and specifications for thermal comfort, including adaptive approaches and energy considerations. It is supplemented by background scientific knowledge, which is presented in ASHRAE Fundamentals.

International Standard ISO 7730 (2005)[2] is the internationally accepted thermal comfort standard and is based mainly on the work of Fanger (1970). Both ASHRAE 55 and ISO 7730 (2005) use similar criteria for thermal comfort and are heavily influenced by the work of Gagge et al. (1971) and Fanger (1970).

---

[2] ISO is the International Standards Organisation of over 150 countries with standards produced by the consensus of experts, a democratic system of voting and a single organisation representing each country. The American National Standards Institute (ANSI) represents the USA, British Standards Institution (BSI) in the UK etc. It is common in the USA for large professional bodies such as ASHRAE to be delegated by ANSI to coordinate the production of standards in their area of expertise. Hence ASHRAE 55 is ANSI/ASHRAE approved. It is a national standard in the USA with influence world-wide.

Professor Fanger's Comfort Equation

2

A PARADIGM SHIFT

P O Fanger was a young engineer from Denmark who presented his doctoral thesis in 1970 on research into thermal comfort conducted over 5 years at the Laboratory of Heating and Air Conditioning, Technical University of Denmark and at the Institute for Environmental Research, Kansas State University where he worked from 1966 to 1967. The doctoral thesis was published as a defining book (Fanger, 1970), which revolutionised thermal comfort research and application. It produced a paradigm shift in an integrated approach that not only demonstrated how to determine comfort conditions but also provided a method for determining, in any environment, the likely thermal sensation, on average, a group of people would experience as well as the percentage of people in the group who would be likely to be dissatisfied. The work has been controversial since its inception and over many years for a number of changing reasons as the subject developed. However, it was timely and fitted well into the rapidly developing computer age. It has inspired research worldwide and has been accepted in national, regional and international standards.

The author initiated this research "...recognizing that the existing knowledge of thermal comfort was quite inadequate and unsuitable for practical application" and that "...the creation of thermal comfort for man is one of the principal aims in environmental engineering and in the entire heating and air-conditioning industry" Fanger (1970).

The origins of the research imply that it is limited in application to air-conditioned buildings, and that was clearly an area of primary interest. However that would not do justice to the final outcome, as the development and formulation of the proposed methods were not restricted in this way and the methods proposed have a more universal scope.

FANGER'S COMFORT EQUATION

Definitions and Area of Application

Thermal comfort is defined as "That condition of mind which expresses satisfaction with the thermal environment" ASHRAE 55-66 (1966). Thermal neutrality is defined as the condition when a person would prefer neither warmer nor cooler surroundings. In certain conditions, such as those where there is asymmetric radiant heat on either side of the body, a person may be in thermal neutrality but not comfortable. For most environments however, this is not the case and Fanger (1970) assumes that thermal comfort and thermal neutrality are the same. Cases where they not, are considered separately as exceptions. In short, if we can predict thermal neutrality, we can predict thermal comfort.

To be clear, Fanger (1970) considers that the main reason for building homes at all is to control the thermal surroundings and that artificial climates include houses, offices, factories, shops, hospitals, schools, theatres, cinemas, meeting rooms, restaurants, hotels, athletic halls, museums, churches, libraries, cars, buses, trams, trains, ships, aeroplanes, space craft and more. In the future, they may include enclosed cities and controlled outdoor environments. The implication is that we require a universal understanding and a method for specifying thermal comfort conditions.

Conditions for Thermal Comfort – Variables and Assumptions

Fanger (1970) begins this endeavour by stating that the most important variables, which influence thermal comfort, are activity level, clothing, air temperature, mean radiant temperature, relative air velocity and water vapour pressure in ambient air. Thermal comfort can be achieved by many different combinations of variables and hence by many technical systems. No one factor can specify thermal comfort, all six factors are required and in combination.

This emphasis was the first revolutionary point made by Fanger who notes that previous studies had tended to consider only a sub-set of possible combinations.

The next important conclusion was that for a person to be in thermal comfort, they must be in heat balance and that their mean skin temperature and sweat secretion must be at levels for thermal comfort. Based upon the above assumptions and a series of reasonable estimations, a comfort equation was derived that enabled the prediction of thermal comfort conditions from the measurement and estimation of values of the six variables.

Heat Balance

The first condition for thermal comfort is that a person is in heat balance. This is a necessary but not sufficient condition for thermal comfort. A person can be hot and uncomfortable but maintain heat balance by sweating or cold and uncomfortable but maintain heat balance by reducing heat loss with low skin temperatures.

Fanger (1970) set up a double heat balance equation for the human body. Some of the symbols and units first used by Fanger have changed over the years and those more recently accepted are used here. As the main objective of human thermoregulation is to maintain an internal body temperature of around 37°C, the body over time must neither have a net gain nor loss of heat. That is there will be a dynamic heat balance as heat production equals heat loss from the body. Heat balance is therefore achieved when heat production in the body (H) minus heat loss at the skin (by water vapour diffusion through the skin (E_d) and the evaporation of sweat from the skin surface (Esw)) and by breathing (latent respiratory heat loss (Eres) and dry respiratory heat loss (Cres)), equals the heat transfer by conduction through clothing (to the surface of the clothing (K)) where it must equal the net heat loss from the surface of clothing (by the sum of radiation (R) and convection (C)). That is $H - E_d - Esw - Eres - Cres = K = R + C$. The challenge was to provide equations for calculating the components of the heat balance equation from values of the six variables that are essential for specifying comfort conditions.

Internal Heat Production (H)

People consume food and 'burn' it in oxygen in their cells to produce energy (the total making up metabolic rate, M). Some of the energy is used for mechanical work (W) but most is converted to internal body heat (H). So $H = M - W$. The units of energy are Joules (J) and rate of energy production or transfer is termed power, where $1 J s^{-1}$ is 1 Watt (W). Fanger (1970) uses kcal per hour (kcal h^{-1})

and attempts to 'normalise' heat production across people of different sizes for a given activity, by dividing the estimate of heat production by an estimate of their surface area. He uses an estimate of surface area for the body from Du Bois and Du Bois (1916) as $A_D = 0.202 W^{0.425} \times H^{0.725}$ where W is the body weight in kilograms and H is the body height in metres. The unit 1 met is a metabolic rate for a person seated at rest and is $50 kcal\ h^{-1}\ m^{-2}$ ($58.15 W\ m^{-2}$). So for a large person at rest with a surface area of $2.0 m^2$ metabolic rate is estimated as 116.3 W, whereas for a small person with a surface area of $1.2 m^2$, metabolic rate would be estimated as 69.8 W. Seated at rest therefore has a normalised value estimated at $58.15 W\ m^{-2}$ for all people.

To obtain an estimate of how much heat is produced in the body for a given activity, expired air can be collected in a (Douglas) bag over a fixed period of time to measure how much oxygen has been used to carry out the task or job (i.e. amount of oxygen in inspired air (around 20%) minus amount of oxygen in expired air (e.g. around 16%)). Using the calorific value of food (heat produced per litre of O_2 used providing about 20 kJ for a normal diet), an estimate of energy and hence heat produced can be made (see Parsons, 2014). This is the method of indirect calorimetry. There are other methods and for all methods the level of accuracy is debatable. Fanger (1970) presents established databases of metabolic rate values for a wide range of activities, including work types and tasks as well as for different walking speeds and gradients. In general, for lying at rest (minimum or basal metabolic rate), values are around $40 W\ m^{-2}$; siting at rest, $58 W\ m^{-2}$; standing relaxed, $70 W\ m^{-2}$; low activity such as slow walking on level ground, $110 W\ m^{-2}$; medium activity such as brisk walking with a 10 kg load, $185 W\ m^{-2}$; and maybe beyond applications involving thermal comfort, high activity such as loading a wheelbarrow, $275 W\ m^{-2}$ and very high activity such as walking uphill or upstairs, $440 W\ m^{-2}$.

Fanger (1970) provides estimated values for mechanical efficiency ($\eta = W/M$) so that heat production can be calculated from metabolic rate ($H = M - W = M (1 - \eta)$). However, these values are speculative and generally considered to be low. In modern considerations, W is often taken as zero and hence $H = M$. Relative air velocity values are also included as movement (e.g. walking) can cause even still air to move across the body and influence heat transfer by both convection and evaporation.

Heat Loss by Skin Diffusion (E_d)

Heat loss by water vapour transfer from the skin is driven by the difference between the concentration of water vapour at the skin and the concentration of water vapour in the air. As the skin is always at least moist, we can assume that there is a saturated vapour pressure at the skin, at skin temperature, ($P_{s,sk}$).

Partial vapour pressure in the air (P_a) is related to humidity $\phi = P_a/P_{sa}$, where P_{sa} is the saturated vapour pressure at air temperature. Heat loss by vapour diffusion through the skin is therefore the permeation coefficient of skin (m) × the latent heat of vaporisation of water (λ) at skin temperature × (the difference between the saturated vapour pressure at skin temperature and the partial vapour pressure in the air). In its modern form the equation presented by Fanger (1970) is $E_d = 3.05 \times 10^{-3}$ (256 t_{sk} − 3,373 − P_a) W m^{-2}, where t_{sk} is mean skin temperature (for comfort and also used in the estimation of $P_{s,sk}$).

Mean Skin Temperature and Sweat Secretion for Comfort

The heat balance equation includes the terms skin temperature (t_{sk}) and the evaporation of sweat at the skin (Esw). The heat balance equation for comfort (the comfort equation) uses skin temperatures and sweat rates that provide comfort. Experiments with human subjects had shown that required evaporation levels for comfort (Esw,req) increased with activity level (active people prefer to sweat more) and that required skin temperatures for comfort (t_{sk},req) decrease with activity level. Best fit regression equations were Esw,req = 0.42 (M − W − 58.15) W m^{-2} and that t_{sk},req = 35.7 − 0.0275 (M − W)°C.

Latent Respiratory Heat Loss (Eres)

A person breaths in air at air temperature that contains water vapour. The respiratory tract and lungs raise the temperature of the air to internal body temperature and saturate the air at that temperature. Expired air therefore transfers heat from the body by evaporation and convection (the combination often termed enthalpy). Actually heat can be transferred into the body in this way but under conditions usually far from comfort. The evaporative (latent) heat loss by breathing is driven by the difference between the amount of water vapour in the inspired air (depends upon humidity) and the amount of vapour in the expired air. Fanger (1970) expresses these as humidity ratios in kg of water per kg of dry air for expired (Wex) and inspired (Wa) air. These can be converted into vapour pressures. Air is saturated at deep body temperature (37°C) in the alveoli of the lungs but as it moves outwards, some heat is transferred back to the body and water is condensed. For comfort conditions, air from the nose and mouth, however, still contain more heat and water than inspired air so breathing causes a dry and latent heat loss from the body. The expired air is assumed to be at 35°C and the heat loss is driven by the difference between the saturated vapour pressure at 35°C and the partial vapour pressure in inspired air (P_a).

How much heat is lost depends upon the amount of air 'conditioned' by the body. This depends upon the pulmonary ventilation ($\dot{V}$ – related to breathing rate and volume of air per breath (normally tidal volume – related to lung capacity)) and the latent heat of vaporisation of water at 35°C ($\lambda = 575\,\text{kcal kg}^{-1}$). That is Eres = $\dot{V}$ (Wex – Wa) λ.

Fanger (1970) reviews the literature and uses a linear approximation between pulmonary ventilation and metabolic rate as $\dot{V} = 0.0060\,\text{M kg h}^{-1}$. This orientates the final comfort equation to using an estimation of metabolic rate rather than using pulmonary ventilation as an input variable. Although useful, it is debatable whether this relationship is valid for all activities from static muscular work to activities involving light but rapid movement. Wex is estimated as 0.029 and Wa as $0.622 \times (P_a/(P - P_a))$, where P is atmospheric pressure. A final equation of therefore Eres = $0.0023\,\text{M}(44 - P_a)\,\text{kcal h}^{-1}$ which converts to Eres = $1.72 \times 10^{-5}\,\text{M}(5{,}867 - P_a)\,\text{W m}^{-2}$ where P_a is in Pascals and M is in W m^{-2}.

Dry Respiratory Heat Loss (Cres)

Inspired air at air temperature is heated by the lungs to body temperature and is eventually expired at a temperature that Fanger (1970) estimates to be tex = 34°C. Heat is therefore lost from the body by convection, and this dry respiratory heat loss (Cres, although Fanger (1970) uses the symbol L) is given as Cres = $\dot{V} \times c_p \times (\text{tex} - t_a)$. The specific heat of dry air at constant pressure (c_p) is given as $0.24\,\text{kcal kg}^{-1}\,°\text{C}^{-1}$ and $\dot{V}$ is estimated as $0.0060\,\text{M kg h}^{-1}$. So Cres = $0.0014\,\text{M}(34 - t_a)\,\text{kcal h}^{-1}$, which converts to a higher multiplying constant as W m^{-2} but interestingly has not been adjusted in future formulations (e.g. ISO 7730, 2005). As this is a relatively small heat transfer and approximations have been used, this has not presented a problem.

Heat Conduction through Clothing (K)

In the double heat balance equation, there is a net transfer of heat from the skin surface to the surface of clothing. In reality this is a complex process involving water vapour and water transfer as well as convection and radiation in air layers and spaces and conduction through clothing fibres. For areas of the body not clothed (e.g. hands, head, etc.) there will be more direct heat transfer and if the body moves, warm air and hence heat will be forced out of clothing through gaps (neck, cuffs, etc.) and 'vents', often part of clothing design. If all of the heat transfer mechanisms are combined into a single index value (a single number equivalent to overall effect), then conduction (K) has units of Watts per °C difference between the mean skin surface temperature (t_{sk}) and the mean surface temperature of clothing (t_{cl}) as if it were evenly distributed across the body.

As clothing restricts heat transfer, this is often represented as a resistance (Rcl) and, when divided by surface area for heat transfer, provides units of m² °C W⁻¹. Gagge et al. (1941) proposed the clo unit and symbol I_{cl} as an expression for the overall thermal resistance from the skin to the outer surface of the clothed body. The clo value is a measure of the overall thermal resistance of clothing across the body and not of the clothing material. It has the units m² °C W⁻¹, where the m² term refers to a square metre of the body surface area. One clo is often said to represent the thermal insulation of a typical business suit and is given a value of I_{cl} = 1.0 clo = 0.155 m² °C W⁻¹. In modern units, Fanger (1970) gives conduction through clothing as K = (t_{sk} − t_{cl})/0.155 I_{cl} W m⁻², where I_{cl} is in clo units.

Fanger (1970) notes that the thickness of clothing is of prime importance in determining thermal resistance and that clo values are measured using a human body-shaped heated thermal manikin. Clothing will increase the surface area available for heat transfer with the environment. That is from A_D for a nude person to the surface area of the clothed body (Acl), the increase represented by f_{cl} = Acl/A_D. Fanger (1970) provides a table of clo values for a wide range of clothing ensembles along with estimates of associated f_{cl} values. These are limited, particularly for civilian clothing and indoor thermal comfort, and it is recommended that further research be conducted, particularly in the area of the water vapour resistance properties of clothing. Much has been done since 1970 (e.g. ISO 9920, 2009, Parsons, 2014) although quantifying the thermal properties of clothing remains complex and a challenge, particularly in extreme environments where people are sweating and active. For thermal comfort applications, Fanger (1970) provides estimates, sufficient for most applications. Examples of clo values for typical clothing ensembles are: for a nude person (0 clo); shorts and short sleeved shirt (0.35 clo); long, light weight trousers (pants) and open neck short sleeved shirt (0.5 clo); cotton work trousers and open neck work shirt (0.6 clo); light outdoor sportswear with light jacket (0.9 clo); typical business suit (1.0 clo); heavy European (woollen) business suit (1.5 clo) and more. Although higher values are provided, it is reasonable to assume that comfort starts to become compromised by clothing weight and volume.

Heat Balance – A Summary So Far

In summary, metabolic heat (H) is transferred from around the body to the skin surface which it is assumed is at a temperature for comfort. In addition, heat is lost by skin diffusion (E_d) and sweating (Esw) assumed to be at an evaporation level for comfort. Heat is also lost by breathing (Eres + Cres). As the skin temperature is higher than the clothing temperature, heat conducts through clothing to the clothing surface. Heat balance is then achieved by a net heat loss from the clothing surface to the environment.

It is worth noting at this point that there is what is called an 'air layer' around the body. Modern considerations define I_a as the insulation of air around a nude body and hence I_a/f_{cl} as the thermal insulation of air around a clothed body as it takes account of the increased surface area for heat exchange due to clothing. I_{cl} as described is termed 'intrinsic clothing insulation'. The total clothing insulation is the intrinsic clothing insulation plus the insulation of the air layer ($I_T = I_{cl} + I_a/f_{cl}$). In theory, I_{cl} is not influenced by environmental conditions whereas I_a and hence I_T are. I_a is often termed the insulation of the environment.

Heat Loss to the Environment by Radiation and Convection (R + C)

Heat Loss by Radiation (R)

Heat transfer by radiation is proportional to the difference between the fourth powers of the absolute temperatures of two surfaces. The constant of proportionality is called the Stefan–Boltzmann constant (σ) and is 5.67×10^{-8} W m^{-2} K^{-4}. Fanger (1970) uses this Stefan–Boltzmann law to express heat transfer by radiation from the outer surface of the clothed body to the environment as R = Aeff ε σ $((t_{cl} + 273)^4 - (t_r + 273)^4)$.

Aeff is the effective radiation area of the clothed body. It takes account of radiation direction and is the surface area of the whole body (A_D) × (the ratio of the effective radiation surface area of the clothed body to the total surface area of the clothed body (feff)) × (the ratio of the surface area of the clothed body to the surface area of the nude body (f_{cl})). That is Aeff = A_D × feff × f_{cl}. Fanger (1970) uses photographic methods to conclude that feff = 0.71 is a good estimate for both sitting and standing postures and for people of different sizes. The emittance of the outer surface of the clothing (ε) is taken as 0.97 for indoor environments out of the sun as a clothed person is similar to a black body for long wavelength indoor radiation conditions. Lower values would apply for short wavelength solar radiation and would depend upon the colour of the clothing. Fanger (1970) therefore gives the following equation for heat transfer by radiation as R = 3.4×10^{-8} A_D f_{cl} $((t_{cl} + 273)^4 - (t_r + 273)^4)$ kcal h^{-1}, which converts to R = 3.96×10^{-8} f_{cl} $((t_{cl} + 273)^4 - (t_r + 273)^4)$ W m^{-2}.

Heat Loss by Convection (C)

Heat transfer by convection involves movement of fluid. In 'still' air, it can be 'natural' (free) due to lighter warm air (next to skin or clothing) rising and being replaced by heavier cooler air due to gravity. It can also be 'forced' by air moving across the body (caused by moving air or human movement or both (e.g. wind, walking)). Natural and forced convection can also act in combination. Fanger (1970) provides the equation for heat transfer by convection as $C = A_D f_{cl} hc (t_{cl} - t_a)$. The convective heat transfer coefficient (hc) is estimated for free convection as $hc = 2.05 (t_{cl} - t_a)^{0.25} kcal\ h^{-1}\ m^{-2}\,°C$, which converts to $hc = 2.38 (t_{cl} - t_a)^{0.25} W\ m^{-2}\,°C$. For forced convection, $hc = 10.4\ v^{1/2} kcal\ h^{-1}\ m^{-2}\,°C$, where v is relative air velocity, which converts to $hc = 12.1\ v^{1/2} W\ m^{-2}\,°C$. The hc value is taken as the larger of the two values (free or forced).

The formulation of Fanger's comfort equation in its modern form (exactly the same equation but with minor change to symbols and units) is provided in Figure 2.1. It contains variables I_{cl} and f_{cl} (for clothing); M/A_D, η and v (related to activity) and t_a, t_r, v and P_a (environmental variables). In modern form, η is replaced by W and all heat transfer and production units are in Watts (m^{-2}).

M – W	H is Metabolic heat production
$- 3.05 \times 10^{-3} [5733 - 6.99(M\text{-}W) - p_a]$	Ed water vapour diffusion through skin at comfortable skin temperature
$- 0.42 \times [(M\text{-}W) - 58.15]$	Esw evaporative heat loss at level for comfort
$- 1.7 \times 10^{-5} M\{ 5867 - p_a)$	Eres evaporative heat loss by respiration
$- 0.0014M (34 - t_a)$	Cres convective heat loss by respiration
$= 3.96 \times 10^{-8} f_{cl} [(t_{cl} + 273)^4 - (t_r + 273)^4]$	R heat transfer by radiation
$+ f_{cl} h_c (t_{cl} - t_a)$	C heat transfer by convection

$$t_{cl} = 35.7 - 0.028\,(M\text{-}W) - 0.155\,I_{cl} [(M\text{-}W) - 3.05 \times 10^{-3} \{5733 - 6.99\,(M\text{-}W) - p_a\} - 0.42\{ (M\text{-}W) - 58.15\} - 1.7 \times 10^{-5}M (5867 - p_a) - 0.0014\,M (34 - t_a)]$$

$$h_c = max\left(2.38\left(t_{cl} - t_a\right)^{0.25}, 12.1\sqrt{v}\right) \quad f_{cl} = 1.0 + 0.2I_{cl}\ or\ f_{cl} = 1.05 + 0.1I_{cl}\ for\ I_{cl} > 0.5$$

FIGURE 2.1 Comfort equation of Fanger (1970) in format of ISO 7730 (2005). $H - E_d - Esw - Eres - Cres = R + C$ (all in W m^{-2}); t_a, air temperature (°C); t_r, mean radiant temperature (°C); t_{cl}, clothing surface temperature (°C); P_a, partial vapour pressure (Pa); v, relative air velocity (m s^{-1}), M, metabolic rate (W m^{-2}); W, mechanical work (W m^{-2}); f_{cl}, increase in surface area due to clothing (ND); I_{cl}, clothing insulation (clo); hc, convective heat transfer coefficient (W m^{-2} °C).

APPLICATION OF THE
COMFORT EQUATION

Using the comfort equation, comfort conditions can be derived for all combinations of the six variables H (=M − W), I_{cl}, t_a, t_r, v and P_a. It can also be shown how one variable can be traded off against another to provide thermal comfort conditions. A computer program to predict the outcome of the equation demonstrated its power and started to answer a multitude of questions about thermal comfort. For example, how humidity trades off against air temperature, how air velocity influences overall comfort especially when moving from still air and natural convection to forced convection and many more. There are millions of combinations and many practical questions that can be investigated.

Fanger (1970) provides the following examples:

1. For rest places and a restaurant at one end of a swimming pool, the comfort temperature for nude (I_{cl} = 0), sedentary persons (H = 58 W m^{-2}) in low relative air velocity (v = 0.2 m s^{-1}) and 70% relative humidity (rh = 70%) where air temperature equals mean radiant temperature ($t_a = t_r$) is 29.1°C as calculated by the comfort equation.
2. Comfort temperatures for a meeting room in winter (I_{cl} = 1.0 clo; v < 0.1 m s^{-1}; rh = 40%; $t_a = t_r$) are 23.3°C. They are 24.9°C in summer when rh = 70% and I_{cl} is 0.6 clo.
3. Optimal temperatures in a clean room with sedentary activity (v = 0.5 m s^{-1}; I_{cl} = 0.75 clo rh = 50% $t_a = t_r$) are found to be 26°C.
4. In a shop where people are walking at 1.5 km h^{-1}, (v = 0.4 m s^{-1}; I_{cl} = 1.0 clo rh = 50%) comfort temperatures are $t_a = t_r$ = 21.0°C.
5. Slaughterhouse workers ($t_a = t_r$ = 8°C; v = 0.3 m s^{-1}; light activity) require clothing insulation of 1.5 clo.
6. In a bus in winter ($t_a - t_r$ = 6°C), air temperature necessary for comfort (I_{cl} = 1.0 clo; v = 0.2 m s^{-1}; rh = 50%) is 26°C.
7. An outside restaurant heated by infra-red heaters to keep air temperature above 12°C (v = 0.3 m s^{-1}; I_{cl} = 1.5 clo; rh = 50%) needs to provide t_r = 40°C for comfort.
8. In a supersonic aircraft ($t_a = t_r − 10°C$), sedentary passengers (I_{cl} = 1.0 clo; v = 0.2 m s^{-1}) should be in air temperatures of 20°C for comfort.
9. In a foundry (t_a = 20°C; t_r = 50°C; rh = 50%; I_{cl} = 0.5 clo; light work (M = 116 W m^{-2})) radiant shields reduce t_r to 35°C. Air velocity should be v = 1.5 m s^{-1} to maintain comfort.

The outputs of all of the above examples and other applications will depend upon the quality and accuracy of the inputs. As with all applications, this will depend upon accuracy of measurement and estimation but also on the formulation and interpretation of the problem. The versatility of the comfort equation is however clear and represented a leap forward in knowledge and application as well as providing direction and stimulating research into thermal comfort.

Fanger (1970) presents curves and diagrams that show the relationship between variables. He also considers validity by comparing predicted thermal comfort conditions with the subjective responses of people. Of particular interest was the comparison with the experimental results of Nevins et al. (1966) and McNall et al. (1968). They found excellent agreement for sedentary subjects but only reasonable agreement for active subjects. The comfort equation can provide predictions for millions of combinations so validation with human subjects can only involve a small sub-set of possible conditions and are therefore 'spot checks'. However, these 'spot checks' usually involve combinations important for practical application so are much more useful than random selections. It is an interesting observation that because of experimental error, the equation (thermal comfort model) may well be more valid than the experimental data. It is certainly more reliable (same inputs give same outputs). However, the use of people who express their satisfaction with the thermal environment is by definition the only way to approach validation. The 'spot checks' provide face validity but the thermal comfort equation (even though not perfect with many assumptions) offers much more and such validation begs the issue of how accurately the thermal conditions have been measured or estimated in empirical studies.

Fanger (1970) considers the influence of certain 'special factors' on the application of the 'comfort equation'. He conducted an extensive literature search as well as carefully designed and controlled experiments to investigate how national geographic location, age, sex, clothing, body build, menstrual cycle, ethnic differences, food, circadian rhythm, thermal transients, asymmetric radiant fields, cold and warm floors, colour of surroundings, crowding, and air pressure, influence comfort conditions. It was concluded that the comfort equation is valid and can be applied in temperate parts of the world, independently of age and sex. He also concludes that body build, menstrual cycle, ethnic differences, food, circadian rhythm, crowding and colour of surroundings do not have influence on thermal comfort conditions which is of practical engineering significance. Other factors have specific influence. They complement the comfort equation in predicting conditions for whole body comfort but require special additional consideration often due to their potential for causing 'local thermal discomfort'.

At this point, Fanger had achieved all of his objectives and had already stirred up considerable controversy. Two major criticisms were that the equation was too (overly) complex for practical application and that people and psychological phenomena such as comfort could not be considered as machines (predictions using engineering principles). The claim that the equation was universal and was valid for all people (across the world) was also challenged. Since 1970 to the present day, there has been substantial research internationally into the claims made for the comfort equation. It remains controversial but it is reasonable to say that many studies, including some of my own, have found the thermal comfort equation to be usable, valid to levels required and robust (Parsons, 2014). Although the equation allows for changes in conditions (different combinations of variables that provide comfort), it does not embrace human behaviour which forms the basis for the next paradigm shift in thermal comfort models.

The Predicted Mean Vote (PMV) and Predicted Percentage of Dissatisfied (PPD)

3

THE PREDICTED MEAN VOTE (PMV)

Fanger did not rest with the comfort equation, which specified conditions for thermal comfort. His next challenge was to establish a practical method of evaluation of thermal environments. No suitable method was available. Starting from the comfort equation, Fanger (1970) derived a new thermal sensation index that predicted the mean sensation vote, on a standard scale, for a group of persons for any given combination of air temperature, mean radiant temperature, air velocity, humidity, activity and clothing. This he called the Predicted Mean Vote (PMV).

Fanger (1970) transposed the commonly used ASHRAE scale (numbered 1, cold; through 4, neutral; to 7, hot) to indicate the PMV from −3 through 0 to +3. That is −3, cold; −2 cool; −1 slightly cool; 0, neutral; +1, slightly warm;

+2, warm; +3, hot. When the comfort equation is satisfied, it would be expected that for a large group of people PMV = 0 (neutral). He argues that as we deviate from optimum conditions (PMV = 0), thermoregulation (vasodilation, vasoconstriction, sweating or shivering) can maintain heat balance even beyond comfort limits. "within these wide limits there is only a small interval which is regarded as comfortable ... we will assume that the thermal sensation at a given activity level is a function of the thermal load L of the body" Fanger (1970). He defined the thermal load on the body as "...the difference between the internal heat production and the heat loss to the actual environment for a man hypothetically kept at the comfort values of the mean skin temperature and the sweat secretion at the actual activity level." The thermal load L = (H − E_d (t_{sk} for comfort) − Esw (for comfort) − Eres − Cres) − (R + C).

In comfort conditions, L will be zero. In other conditions, thermoregulatory responses attempt to achieve heat balance and hence skin temperatures and sweat secretions deviate from comfort conditions. The thermal load is therefore related to the physiological strain on the body. How it relates to thermal sensation, expressed by a mean subjective vote (Y) is determined by subjective responses of people. Fanger (1970) speculated that this relationship might depend upon heat production within the body.

The results of studies by Nevins et al. (1966), McNall et al. (1968) and Fanger (1970), covering the responses of a total of 1,396 subjects clothed in the Kansas State University (KSU) standard uniform (0.6 clo) and exposed for 3 hours to controlled conditions, were used to estimate the relationship between ambient temperature and L. The data included responses involving four levels of activity (sedentary, low, medium, high). The relationship between mean vote (Y) and L was then determined.

Using graphical methods, the rate of change of thermal sensation (Y) with thermal load (L) was determined for Y = 0 (mean thermal sensation of neutral) as around comfort was the area where the rate of change was of most interest. It was found that the rate of change ($\delta Y/\delta L$) was higher for sedentary activity than for higher activity levels and concluded that Y = (0.352exp$^{-0.042(M/A_D)}$ + 0.032) × L was a good fit to the data. This was converted to Y = PMV = (0.303exp$^{-0.036(M)}$ + 0.0275) × L for M in W m^{-2} by fixing the curve to the sedentary activity point on the graph for which there was much more experimental data. The full equation is provided in Figure 3.1. Example calculated PMV values are provided in Table 3.1.

Due to a lack of data for conditions away from sedentary comfort, the predictions on the warm and cold side of comfort are logical and reasonable but more speculative. Rather than a continuous curve relating Y and L, the data suggest two categories (sedentary and active) as once active the mean sensations do not appear to change with the level of activity but only the load. McIntyre (1980) notes that the PMV had not been validated over a range of

$$PMV = \left(0.303e^{-0.036M} + 0.028\right) \times \left[\left(M - W\right)\right.$$

$$-3.05 \times 10^{-3} \left\{5733 - 6.99\left(M - W\right) - P_a\right\} - 0.42\left\{\left(M - W\right) - 58.15\right\}$$

$$-1.7 \times 10^{-5} M\left(5867 - P_a\right) - 0.0014\, M\left(34 - t_a\right)$$

$$-3.96 \times 10^{-8} f_{cl}\left[\left(t_{cl} + 273\right)^4 - \left(t_r + 273\right)^4\right] - f_{cl}h_c\left(t_{cl} - t_a\right)\right]$$

$$t_{cl} = 35.7 - 0.028\left(M - W\right)$$

$$-0.155\, I_{cl}\left[3.96 \times 10^{-8} f_{cl}\left\{\left(t_{cl} + 273\right)^4 - \left(t_r + 273\right)^4\right\} + f_{cl}h_c\left(t_{cl} - t_a\right)\right]$$

$$h_c = \max\left(2.38\left(t_{cl} - t_a\right)^{0.25}, 12.1\sqrt{v}\right) \quad \text{fcl} = 1.0 + 0.2\text{Icl} \quad \text{or} \quad \text{fcl} = 1.05 + 0.1\text{Icl for Icl} > 0.5$$

FIGURE 3.1 Equation for calculating the Predicted Mean Vote (PMV) from Fanger (1970). In the format of ISO 7730 (2005). t_a, air temperature (°C); t_r, mean radiant temperature (°C); t_{cl}, clothing surface temperature (°C); P_a, partial vapour pressure (Pa); v, relative air velocity (m s^{-1}); M, metabolic rate (W m^{-2}); W, mechanical work (W m^{-2}); f_{cl}, increase in surface area due to clothing (ND); I_{cl}, clothing insulation (clo); hc, convective heat transfer coefficient (W m^{-2}°C).

clothing and activity levels but gives good results, as expected, for sedentary activity and light clothing. He also notes that responses of only male subjects were used in the derivation as those of female subjects were more scattered. He is not convinced by the central concept of thermal load and gives the example that a man in heavy clothing would experience a greater increase in skin and body temperature than a man in light clothing for the same increase in metabolic rate and have a greater increase in thermal sensation. McIntyre (1980) presents a simplification of the PMV equation for ease of calculation but this has not been adopted and the argument of complexity of calculation has been swept away by the computer age. This also made redundant the extensive charts and tables provided by Fanger (1970).

THE PREDICTED PERCENTAGE OF DISSATISFIED (PPD)

The PMV predicts the mean vote of a large group of people exposed to a given combination of the variables. Although it indicates when the environment is optimum (PMV = 0, Neutral), and the degree of discomfort away from neutral, it does not indicate how much dissatisfaction there will be among the group

TABLE 3.1 PMV values for typical indoor environments for two levels of activity (1.0 and 1.2 mets), low air movement, 50% relative humidity, three levels of clothing (0.5, 0.75 and 1.0 clo) and air temperature = mean radiant temperature

MET.	1.0			1.2		
$t_a = t_r/I_{cl}$	0.5	0.75	1.0	0.5	0.75	1.0
18	999	−2.27	−1.58	−2.17	−1.42	−0.88
19	−2.88	−1.96	−1.30	−1.92	−1.19	−0.68
20	−2.51	−1.64	−1.02	−1.58	−0.92	−0.45
21	−2.13	−1.30	−0.74	−1.31	−0.68	−0.23
22	−1.75	−0.98	−0.46	−0.97	−0.40	0.00
23	−1.33	−0.66	−0.18	−0.69	−0.15	0.23
24	−0.95	−0.33	0.10	−0.36	0.12	0.46
25	−0.56	−0.01	0.38	−0.06	0.37	0.69
26	−0.18	0.31	0.66	0.26	0.64	0.91
27	0.20	0.64	0.95	0.57	0.91	1.15
28	0.59	0.96	1.24	0.88	1.18	1.37
29	0.98	1.31	1.54	1.22	1.46	1.63
30	1.37	1.65	1.83	1.51	1.70	1.83
31	1.80	1.99	2.13	1.87	2.01	2.11
32	2.21	2.34	2.43	2.17	2.29	2.36
33	2.62	2.69	2.74	2.54	2.57	2.60

Relative air velocity = 0.15 m s⁻¹, Relative humidity = 50%.
Values correspond to the scale 3, hot; 2, warm; 1, slightly warm; 0, neutral; −1, slightly cool; −2,cool; −3, cold. (e.g. for the conditions specified, at rest (1.0 met) in a business suit (1.0 clo) would give a PMV value of −1.02 at 20°C corresponding to slightly cool).
999 indicates off scale.

for a given mean sensation vote. The Predicted Percentage of Dissatisfied (PPD) predicts the number of 'potential complainers' or persons expected to be 'decidedly dissatisfied' or 'decidedly uncomfortable' and therefore dissatisfied. As people are not alike, there will be a variation in dissatisfaction across the group. Fanger (1970) uses the studies from the USA and Denmark, involving 1,296 subjects who were exposed for 3 hours to a given ambient temperature while wearing light clothing (0.6 clo). They provided votes of thermal sensation every 30 minutes throughout 3 hours and the mean of the last three votes for each subject was 'reckoned' to the nearest vote (integer). The distribution of the mean votes at each ambient temperature from 66°F (18.9°C) to 90°F (32.2°C) was then derived.

The dissatisfied were taken as those who voted −3 (cold); −2 (cool); +2 (warm); +3 (hot). Using probit analysis, the proportion of individuals thermally dissatisfied over the range of ambient temperatures investigated was plotted. Two lines were derived, one for cold dissatisfied and one for warm dissatisfied. This was then converted from ambient temperature to the PMV values, derived from the controlled experimental conditions, and presented, in a semi-logarithmic plot, as the total percentage of dissatisfied as a function of the PMV. This was provided in graphical form but an equation is provided in later publications as PPD = $100 - 95\exp(-0.03353 \text{ PMV}^4 - 0.2179 \text{ PMV}^2)$. It can be seen that the PPD is symmetrical about PMV = 0 and that because individuals are different that the minimum PPD is 5% (not everyone can be satisfied). For PMV = ±1, PPD is 26.8%; PMV = ±1.5, PPD is 52.0%; PMV = ±2, PPD is 76.4%; (or 25%, 50% and 75% in round numbers). When PMV is negative the majority of the dissatisfied will be dissatisfied due to cold, and when PMV is positive, most will be dissatisfied due to heat. There will be equal numbers of dissatisfied due to cold and due to heat at PMV = 0. At PMV = +0.8 (neutral to slightly warm) of the 18.5% of people dissatisfied, 0.1% of people are dissatisfied due to cold and at PMV = −0.8 (neutral to slightly cool) of the 18.8% of people dissatisfied, 0.1% due to heat. Fanger (1970) concludes that at PMV > ±1.0, and hence away from comfort, the dissatisfied will make up a homogeneous group leading to a 'massive demand for unilateral change'.

The PPD provides a method for specifying 'acceptable conditions'. A range of conditions where PPD is less than 7.5% (PMV within ±0.35) is given as an example by Fanger (1970). More recent suggestions have been for gold, silver and bronze quality spaces (sometimes labelled A, B and C) based on PMV = ±0.2, ±0.5 and ±7, respectively (ASHRAE 55, 2017; EN 15251, 2007; ISO 7730, 2005).

THE LOWEST POSSIBLE PERCENTAGE OF DISSATISFIED (LPPD)

Not only are there individual differences in the thermal sensitivity of people (and possibly the same individual over time) but they will be exposed to different conditions across a space (e.g. a room). Each person occupies a different space. If we assume that each person in the space is exposed to identical conditions, as determined by the six essential variables, then the space is homogeneous and the corresponding PMV and PPD values can be calculated. If the PMV value across the room is zero, then the Lowest Possible Percentage of Dissatisfied (LPPD) will be 5%. If the conditions across the space (room) vary, then PMV 'comfort' contours can be provided on a plan of the room as well as corresponding PPD contours showing where complaints are likely to come from. It follows that if the average conditions are adjusted (by varying one or a combination of the variables) to provide an average PMV of zero, this will provide the LPPD for that space. The LPPD therefore provides an index of non-uniformity across the room. Fanger (1970) focusses on heating and air conditioning systems in rooms and adjustments to air temperature. In theory, any of the six, or any combination of, the six variables can be involved in producing an average PMV of zero (comfort). For example, changing levels of clothing, using fans and adjusting air temperature. Fanger (1970) suggests that near to comfort values, changing air temperature should not influence other factors such as air velocity.

Fanger (1970) suggests a target value of LPPD < 6% and provides advice for those involved in environmental design, environmental engineering and building services, for example, to take into account mean outdoor 24-hour temperature under clear sky conditions. If the LPPD is greater than 'specified acceptable' levels, then attention must be paid to individual areas of the room. This involves overall environmental design considerations. The LPPD also contributes to thermal comfort survey methods for the assessment and evaluation of spaces (see Chapter 11).

LIMITATIONS OF THE PMV AND PPD THERMAL COMFORT INDICES AND THE NEXT PARADIGM SHIFT

The PMV and PPD indices remain controversial but have been widely adopted internationally (e.g. ISO 7730, 2005). The PMV index can be regarded as a useful starting point but it has limitations as it does not account for human behaviour. A subtle point is that it does account for the consequences of human behaviour as if the environment to which people are exposed to change from one set of (steady state) conditions to another, the PMV will change. However, the PMV does not account for the disposition that if a person is uncomfortable, they will do something about it. If a person is cold, they will attempt to 'warm up'. If they are hot, they will attempt to 'cool down'. Good environmental design will provide opportunities to do so. The PMV is 'passive', people are active.

Professor Pharo Gagge's New Effective Temperature (ET*)

4

THERMAL COMFORT USING ET* AND THE SET

It is a reasonable generalization to say that while research in the UK concentrated mainly on practical solutions to providing thermal comfort, from the 1960s onwards more universal solutions were pursued. This was mainly at the Technical University of Denmark and in the USA, at Kansas State University and the J B Pierce Foundation laboratory, Yale University, often supported by the American Society of Heating, Refrigerating and Air-conditioning Engineers (ASHRAE).

Pharo Gagge conducted research in the 1930s and was influential in the development of the concepts of the clo (Gagge, 1937) unit as well as skin wettedness (w) (Gagge et al., 1941). The clo unit was initially used to indicate to 'senior' military personnel how much clothing was required for soldiers performing a range of activities and operating in a wide range of climates. It has since become a universally accepted unit in considerations of human heat stress, cold stress and thermal comfort. The skin wettedness (w) is defined as the actual amount of evaporation from the skin surface (Esk) divided by the maximum evaporation that could occur in that environment (Emax). It therefore has a theoretical value from 0 to 1. The minimum value is often taken as w = 0.06 for insensible perspiration (no sweating but the body is always moist to some extent) and a maximum value of w = 1.0, where heat lost by evaporation from the body (skin) is at the maximum it can be (influenced by humidity, air velocity and clothing). Skin wettedness can be regarded as a useful index of heat stress, and Gagge et al. (1971) demonstrate that it can also be a valid indicator of discomfort during regulatory sweating. It is argued that skin wettedness is a more valid indicator of discomfort than sweat rate, which was proposed by Fanger (1970). An example is given that if humidity changes (and other variables remain the same), then the sweat rate will not change as 'sensible' (dry) heat loss and metabolic heat remain unchanged so sweat rate remains at the level to maintain heat balance. However, as humidity increases, the maximum possible evaporation decreases and as the actual evaporation will remain at that which is required to maintain heat balance, skin wettedness will increase. This explains why an increase in humidity makes a person feel warmer even when the sweat rate does not change. Just as Fanger (1970) defined comfort conditions as a person in heat balance with sweat rates and skin temperatures at levels for comfort, so Gagge et al. (1971) defined equivalent (comfort and other) conditions as having the same heat loss at the skin and the same skin temperatures and the same skin wettedness.

Gagge et al. (1971) defined the New Effective Temperature (ET*) as the temperature of a standard environment ($t_a = t_r$; $v < 0.15\,m\,s^{-1}$; rh = 50%) in which a person would have the same heat loss, at the same mean skin temperature and the same skin wettedness as he *(or she)* does in the actual environment. The ET* was derived for use in the region of thermal comfort conditions for people in rooms conducting light work in light clothing. It was a development of the original Effective Temperature (ET) index of Houghton and Yagloglou (1923) providing a more realistic standard environment (the ET uses rh = 100%) and in ET* equivalence is rationally defined rather than being based upon subjective judgements.

THE STANDARD EFFECTIVE TEMPERATURE (SET)

Gagge et al. (1973) extended the ET* to provide a more universal index that could be used for considerations of heat stress, cold stress and thermal comfort. This is the Standard Effective Temperature (SET) that is defined as the temperature of an isothermal room ($t_a = t_r$; still air; rh = 50%) in which a person with a standard level of clothing insulation would have the same heat loss, the same mean skin temperature and at the same skin wettedness as he does in the actual environment and clothing insulation under consideration. As with the work of Fanger (1970), it was recognised that the condition of the body for thermal comfort (or any sensation) will depend upon the activity level. So that the same SET value provided neutral comfort (24°C, SET), the standard level of clothing used in the definition of the standard environment is modified as a function of activity. So for 1.1 Mets (1 Met = 58.15 W m^{-2}) activity, the standard clothing is 0.6 clo, mean body temperature (a weighted average of mean skin temperature and core temperature) t_b = 36.35°C and skin wettedness is w = 0.07. For 2.9 Mets, the standard clothing is 0.4 clo, t_b = 36.71°C, w = 0.21. The SET is an extension of the ET*, so that it can be used to assess not only moderate environments but also hot and cold environments. As the standard clothing varies with activity level, the SET value can be directly related to sensation and does not depend upon metabolic rate.

The SET is a direct precursor to the Universal Thermal Climate Index (UTCI), which was derived by a team of international experts, mainly for outdoor environments, to be linked to weather forecasts (Jendritzky et al., 2012). The ET* has continued to be adopted by ASHRAE thermal comfort standards since 1974 (ASHRAE 55-74) where comfort conditions are presented on psychrometric charts. This method of presentation, as well as the use of a standard environment, complement methods employed in the heating and air-conditioning industry. It is different from, but not incompatible with, the Predicted Mean Vote (PMV)/Predicted Percentage of Dissatisfied (PPD) method of Fanger (1970) where it can be concluded that thermal sensation will be the same if the PMV (made up of different combinations of the six variables) is the same. For example, if a combination of the six variables (t_a = 22°C; t_r = 26°C; I_{cl} = 1.0 clo; v = 0.2 m s^{-1}; (M − W) = 1 Met (58.15 W m^{-2}); rh = 70%) provides PMV = 0 (neutral and comfortable), and PMV = 0 is also achieved by (t_a = t_r = 24°C; I_{cl} = 0.6 clo; v = 0.15 m s^{-1}; (M − W) = 1.0 Met (58.15 W m^{-2}); rh = 50%), which is the SET standard environment, then the index value derived from the PMV is 24°C. This will not be identical to the SET as different methods are used to determine equivalence, however, particularly near to comfort, there is little difference between the two indices of practical significance. In later years, ASHRAE effectively adopted both the ET* and the PMV in the specification of thermal comfort conditions. It also follows that just as SET can be estimated using the PMV equation, so can the PMV be estimated using the SET index by similar procedure to that described above.

DETERMINATION OF THE NEW EFFECTIVE TEMPERATURE (ET*) AND STANDARD EFFECTIVE TEMPERATURE (SET)

The ET* represents a sub-set of the conditions covered by the SET and both require a determination of the temperature of the standard environment that would provide the same heat loss at the skin (Hsk), the same mean skin temperature (tsk) and the same skin wettedness (w) as in the actual environment. For the original ET, equivalence was determined by people walking between rooms and making subjective judgements. For the ET* and the SET, in practice, equivalence is determined using a thermal model, in particular, Gagge's two-node model of human thermoregulation.

We should reflect at this point that our problem is complex and any solution only achievable by approximation sufficient for application. This is generally true for scientific research but particularly true for research involving people. Each of the six variables by which we define the thermal conditions and consequent human responses are in continuous and dynamic interaction that change with time. For the purposes of calculation and prediction, the (mostly reasonable) assumption made is that representative values of the six variables (parameters) can be used. The heat loss at the skin, mean skin temperature and skin wettedness are then predicted over time (using the two-node model), starting from comfort conditions and with values predicted at 1 hour of exposure taken to determine the index values (ET* and SET).

THE TWO-NODE MODEL

Passive and Controlling Systems

Gagge's two-node model of human thermoregulation is a 'lumped parameter' mathematical (computer) model made up of two concentric cylinders (inner core and outer shell) attributed to the (estimated average) thermal properties of the human body. This is called the 'passive' system of the model. The active 'controlling' system of the model responds to the conditions of the passive system (temperatures) and varies the relative proportion (size) of shell and core cylinders, simulating vasodilation (large core and reduced shell) and vasoconstriction (large shell and reduced core) and changes in mean skin temperatures as well as heat transfer by evaporation (simulating sweating) and hence predicted skin wettedness. Metabolic rate can also be increased (simulating shivering). The responses are controlled by comparison of temperatures with 'set-point' temperatures, the differences of which drive the different controlling thermoregulatory actions. Details are provided by Gagge et al. (1971), Nishi and Gagge (1977), McIntyre (1980), Parsons (2014, 2019). The dynamics of the model are driven by the conditions of the shell and core at any point in time and the heat transfer to and from the body (passive system) determined by heat transfer equations including the resistance of heat transfer due to clothing.

Starting Conditions (t = 0)

To 'run' the model, it requires a specification of the passive and controlling systems as initial (starting) conditions (time t = 0). These could represent a person from hot to cold including thermal comfort. It also demonstrates that the model can simulate changing conditions as the end of the first exposure to one set of conditions can be used as the start of the next exposure for different conditions. Heat transfer equations are also required. For thermal comfort and the determination of the ET* and SET, comfort conditions of core temperature = 36.8°C; shell (skin) temperature = 33.7°C and a skin wettedness = 0.06 are used as starting conditions. Mean body temperature is a weighted average of core and skin temperature and for comfort $t_b = 0.1 \times 33.7 + 0.9 \times 36.8 = 36.49°C$. The inputs to the model are t_a, t_r, v, rh, I_{cl} and M. There are many possible outputs but those required for determination of the comfort indices are heat loss from the body (Hsk), mean skin temperature (tsk) and skin wettedness (w) after 1 hour of exposure. The model is used to determine these values for the actual environment and again repeatedly with the conditions of the standard environment but systematically varying the air temperature until the same Hsk, tsk and w as in the actual environment occurs. This air temperature is the SET (which for comfort conditions is also the ET*).

How the Model Works

From the initial starting conditions, the input values are used to calculate the heat transfer between the body and the environment and to determine overall heat storage in the body. Negative heat storage indicates moving from comfort to cold and positive heat storage from comfort to hot. The thermoregulatory (controlling) system then responds to restore heat balance (zero heat storage) by adjusting the conditions of the body (mainly the skin) to provide appropriate heat loss or gain by dry (radiation and convection) and evaporative heat transfer. That is such that heat balance is restored and core temperature is maintained at 36.8°C. In summary, if there is a heat gain, then core size increases relative to shell (simulating vasodilation and an increase in skin temperature) and evaporation increases (simulating sweating and leading to an increase in skin wettedness). If there is a heat loss, then shell size increases relative to core size (simulating vasoconstriction leading to a fall in skin temperature) and there may be an increase in metabolic heat production (simulating shivering). The controlling responses are triggered by 'set points' involving skin temperature, core temperature and mean body temperature. The heat storage and responses are determined over a fixed period of time. This is often taken as 1 minute. The change in the body from t = 0 to t = 1 (minute) is then determined. The next starting point is t = 1 (with new starting conditions at t = 1 including Hsk, tsk, w, etc.) and the new conditions at t = 2 are then determined as described above. After 60 minutes (1 hour) the Hsk, tsk and w are recorded and SET (ET*) determined. By presenting the values every minute, (or, for convenience, less frequently for longer exposures – every 15 minutes, every 1 hour for an 8-hour exposure, etc.), a simulation of how the body will respond with time can be made. For comfort conditions and within the maximum capacity of the physiological responses and thermoregulatory controllers, a steady state will be reached, where Hsk, tsk and w will not vary after that state is reached. SET values and predicted corresponding physiological and thermal sensation responses are provided in Table 4.1.

TABLE 4.1 Standard effective temperatures (SET) and corresponding thermal sensation and physiological response

SET (°C)	SCALE VALUE	SENSATION	MEAN BODY TEMPERATURE (t_b) (°C)	PHYSIOLOGICAL STATE, SEDENTARY PERSON
>37.5	>3	Very hot, very uncomfortable	>36.7	Failure of regulation
34.5–37.5	2–3	Hot, very unacceptable	36.8	Profuse sweating
30.0–34.5	1–2	Warm, uncomfortable, unacceptable	36.6	Sweating
25.6–30.0	0.5–1	Slightly warm, Slightly unacceptable	36.5	Slight sweating, accepted vasodilation
22.2–25.6	–0.5 to 0.5	Comfortable and acceptable	36.3	Neutrality
17.5–22.2	–1 to –0.5	Slightly cool, slightly unacceptable	35.8	Vasoconstriction
14.5–17.5	–2 to –1	Cool and unacceptable	35.0	Slow body cooling
10.0–14.5	–3 to –2	Cold, very unacceptable	34.3	Shivering

COMPARISON OF PMV AND SET THERMAL COMFORT INDICES

Fanger (1970) demonstrated the utility of his comfort equation using a diverse range of practical examples (see Chapter 2). These are used below to compare the PMV and SET indices at thermal neutrality and hence comfort conditions (i.e PMV = 0 and SET = 24°C). The comfort equation is represented by combinations of variables in graphical form by Fanger (1970). The 'actual' PMV values are calculated here from a computer program. The PMV values are calculated for the actual conditions in the example to confirm PMV = 0 and for the standard environmental conditions (using actual metabolic rate and clothing adjusted for metabolic rate) of SET = 24°C. The physiological responses for the actual and standard conditions are compared using the two-node model and the temperatures of the standard environments (SET) are calculated for PMV = 0. The corresponding predicted comfort responses are compared with criteria provided in Table 4.1.

1. For rest places and a restaurant at one end of a swimming pool, the comfort temperature for nude (I_{cl} = 0), sedentary persons (H = 58 W m^{-2}) in low relative air velocity (v = 0.2 m s^{-1}) and 70% relative humidity (rh = 70%) where air temperature equals mean radiant temperature (t_a = t_r) is 29.1°C as derived from the comfort equation (Fanger, 1970). For these conditions, values calculated using the computer program (see Chapter 10) are PMV = −0.13 which is close to 'neutral' as predicted. It also shows physiological responses under those conditions (from the two-node model) as core temperature maintained at tcr = 36.83°C indicating heat balance; mean skin temperature at a comfortable tsk = 33.94°C and skin wettedness at a not uncomfortable w = 0.09. Mean body temperature is t_b = 36.53°C which is around SET values for comfort. Responses for the SET standard environment for 24°C (i.e t_a = t_r = 24°C; v = 0.15 m s^{-1}; actual metabolic rate = 1 Met = 58.15 W m^{-2}; standard clothing at 1 Met = 0.63 clo and 50% rh) are predicted as tcr = 36.81°C; tsk = 33.44°C; w = 0.06 and t_b = 36.4°C. That is very similar to those predicted in the actual environment and within the comfort range. So the use of the PMV and SET would give similar interpretations.

2. Comfort temperatures for a meeting room in winter (I_{cl} = 1.0 clo; v < 0.1 m s^{-1}; rh = 40%; t_a = t_r) are 23.3°C. Calculated PMV values are close to 'neutral' with PMV = −0.18 with predicted physiological responses around comfort as tcr = 36.82°C; tsk = 33.94°C; w = 0.08;

$t_b = 36.52°C$. For those conditions the SET would be around 24°C and hence comfortable as also indicated by the PMV.

3. Comfort temperatures for a meeting room in summer ($I_{cl} = 0.6$ clo; $v < 0.1$ m s^{-1}; rh = 70%; $t_a = t_r$) are 24.9°C. Calculated PMV values are close to 'neutral' with PMV = −0.21 with predicted physiological responses around comfort as tcr = 36.82°C; tsk = 33.93°C; w = 0.10; $t_b = 36.52°C$. For SET = 25°C, almost identical physiological responses were predicted by the two-node model for standard conditions. This would be within comfortable conditions as also indicated by the PMV values.

4. Optimal temperatures in a clean room with sedentary activity ($v = 0.5$ m s^{-1}; $I_{cl} = 0.75$ clo; rh = 50%; $t_a = t_r$) are found to be 26°C. Calculated PMV values are close to 'neutral' with PMV = −0.09 with predicted physiological responses around comfort as tcr = 36.82°C; tsk = 33.90°C; w = 0.07; $t_b = 36.51°C$. For SET = 25°C, almost identical physiological responses were predicted by the two-node model for standard conditions. This is within comfortable range for SET (Table 4.1).

5. In a shop where people are walking at 1.5 km hour^{-1}, ($v = 0.4$ m s^{-1}; $I_{cl} = 1.0$ clo; rh = 50%) comfort temperatures are $t_a = t_r = 21.0°C$. Calculated PMV values are close to 'neutral' with PMV = −0.05 with predicted physiological responses around comfort as tcr = 36.86°C; tsk = 33.89°C; w = 0.14; $t_b = 36.62°C$. For SET = 23.5°C, almost identical physiological responses were predicted by the two-node model for standard conditions which is within the comfortable range.

6. Slaughterhouse workers ($t_a = t_r = 8°C$; $v = 0.3$ m s^{-1}; medium to high activity) require clothing insulation of 1.5 clo. Calculated PMV values are close to 'neutral' with PMV = 0.06 and predicted physiological responses for these values as tcr = 36.94°C; tsk = 34.07°C; w = 0.2; $t_b = 36.76°C$. For SET = 20°C, similar physiological responses were predicted by the two-node model for standard conditions as tcr = 36.95°C; tsk = 34.07°C; w = 0.27; $t_b = 36.77°C$. The SET interpretations would be that the workers would be slightly cool whereas the PMV indicates comfort.

7. In a bus in winter ($t_a − t_r = 6°C$), air temperature necessary for comfort (Icl = 1.0 clo; $v = 0.2$ m s^{-1}; rh = 50%) is 26°C. Calculated PMV values are close to 'neutral' with PMV = −0.08 with predicted physiological responses for those conditions as tcr = 36.82°C; tsk = 33.79°C; w = 0.06; $t_b = 36.49°C$. For SET = 24.4°C, almost identical physiological responses were predicted by the two-node model for standard conditions. This is within comfortable conditions indicated by both SET and PMV.

8. An outside restaurant heated by infra-red heaters to keep air temperature above 12°C (v = 0.3 m s⁻¹; I_{cl} = 1.5 clo) need to provide t_r = 40°C for comfort. Calculated PMV values are close to 'neutral' with PMV = 0.15 with predicted physiological responses for those conditions as tcr = 36.83°C; tsk = 34.37°C; w = 0.12; t_b = 36.59°C. For SET = 27°C, almost identical physiological responses were predicted by the two-node model for standard conditions. This is slightly warm and slightly unacceptable whereas the PMV indicated comfort.

9. In a supersonic aircraft (t_a = t_r − 10°C), sedentary passengers (I_{cl} = 1.0 clo; v = 0.2 m s⁻¹) should be in air temperatures of 20°C for comfort. Calculated PMV values are close to 'neutral' with PMV = −0.18 with predicted physiological responses around comfort as tcr = 36.82°C; tsk = 33.9°C; w = 0.07; t_b = 36.51°C. For SET = 25°C, almost identical physiological responses were predicted by the two-node model for standard conditions. This implies that the conditions would be indicated as comfortable by both SET and PMV.

10. In a foundry (t_a = 20°C; t_r = 50°C; rh = 50%; Icl = 0.5 clo; light work (M = 116 W m⁻²)), radiant shields reduce t_r to 35°C. Air velocity should be v = 1.5 m s⁻¹ to maintain comfort. Calculated PMV values are close to 'neutral' with PMV = −0.10 with predicted physiological responses around comfort as tcr = 36.91°C; tsk = 34.01°C; w = 0.18; t_b = 36.71°C. For SET = 22°C, similar physiological responses were predicted by the two-node model for standard conditions. When PMV indicates comfort therefore SET indicates comfortable towards slightly cool.

As noted in Chapter 2, the outputs of all of the above examples and other applications will depend upon the quality and accuracy of the inputs. As with all applications this will depend upon accuracy of measurement and estimation but also on the formulation and interpretation of the problem. From the results, it can be seen that the use of the SET (ET*) and PMV indices provides similar predictions and specifications for thermal comfort that would lead to almost identical practical outcomes for people conducting sedentary or light activity and 'normal' clothing. For higher levels of activity and clothing, the results seem to diverge as the SET seems to be more sensitive to activity in particular than the PMV. This demonstrates a major flaw in the predictive powers of what are called rational indices (analytical methods based upon heat transfer). That is their outcomes are highly sensitive to estimates of clothing insulation and particularly of metabolic heat production due to activity. In short, the variables to which the indices are most sensitive are the least reliable estimates. I concluded this in one of my earliest studies of thermal comfort (Parsons and

Clark, 1984). This included an evaluation of different convective heat transfer coefficients for use in the PMV equation as there were more up-to-date versions. Using different convective heat transfer, coefficients had some effect but was swamped by the estimation of metabolic heat production. So it was concluded that outcomes of rational models very much depended upon estimates of clothing and metabolic rate, but these estimates could be in significant error.

Predicting thermal comfort conditions from rational models with any accuracy would appear to be problematic. We should not however underestimate the face validity and in particular the reliability of such models. If used throughout the world the same inputs will give the same outputs and they include, and are influenced by, all relevant variables that affect thermal comfort in people. They have great value for standardisation for those reasons. As we move towards including human behaviour and adaptive methods into future models of thermal comfort, predicting clothing properties and metabolic heat production will become even more complex as they will become dynamic as a person adapts to the environment. This may not be a great problem, however, as ranges of conditions will be specified in future, where people will take adaptive opportunities to adapt within a range of conditions to achieve thermal comfort. Accurate estimates of specific variables may not then be needed as a range of conditions will allow a range of estimates to be specified.

The present position of providing insulation, vapour permeation and other properties of clothing, usually measured on a sophisticated thermal manikin, is of great value and useful for specification. The ability to adjust and remove clothing in adaptive behaviour, however, provides the need to be able to estimate clothing properties in terms of ranges depending upon how they are adjusted and worn. A business suit with a jacket that can be opened may allow a range of 0.7–1.0 clo, whereas a suit that is not adjusted will have an insulation of 1.0 clo. A challenge for the future will be to measure and specify the thermal properties of clothing over their ranges of adjustment.

Local Thermal Discomfort

5

'NEUTRAL' BUT UNCOMFORTABLE

It can be assumed that a person with an overall thermal sensation of 'neutral' prefers to be no warmer or no cooler. If they prefer to be warmer, it implies that they are uncomfortably cool; and if they prefer to be cooler, it implies that they are uncomfortably warm. A sensation of neutral therefore implies satisfaction, 'no change' and that the person can be assumed to be in 'whole body' or overall, thermal comfort. There are, however, conditions where a person may indicate that overall, they have a neutral sensation, but that their thermal sensation varies over different parts of the body and they are not comfortable. For example, neutral overall, but have uncomfortably warm hands and uncomfortably cold feet or they may be warm to the left side of the body and cold to the right side of the body. A rating of 'neutral' therefore does not indicate comfort as the person is experiencing local thermal discomfort.

It follows that away from neutral, a person may be hot overall but have thermal discomfort caused by cold feet, or they can be cold overall but have discomfort due to a hot head. The causes of local thermal discomfort are usually described as draughts (unwanted convective cooling of the skin caused by air movement), local thermal radiation and radiant asymmetry, vertical temperature gradients and contact with hot or cold objects and surfaces (liquids, handrails, hot or cold floors, etc.). Whatever the cause, it is the sensation caused by a raised or lowered skin temperature that determines the local thermal discomfort. For a person to be in thermal comfort therefore, the body should be in heat balance, mean skin temperature and sweating (wettedness) should be at levels for comfort and there should be no local thermal discomfort.

BODY MAPS, LOCAL THERMAL SENSATION AND SWEATING

If the body becomes too hot or too cold, it is detected by sensors in the skin, transmitted to the brain (via the spinal cord) and effector mechanisms from the hypothalamus (or directly from the spinal cord) stimulate a change in skin condition. This occurs for the whole body but the 'shortcut' via the spinal cord can cause a local reaction. If cold, the body will reduce blood flow to the skin, if hot the body will increase blood flow to the skin and begin sweating if required.

The skin contains nerve endings some of which respond to cold and others heat (there are also others which respond to pressure and others that produce pain). They are distributed differently across the whole surface area of the body. The area stimulated, position and duration of the stimulus, pre-existing thermal state, temperature and rate of change of temperature all affect thermal sensation.

Stevens et al. (1974) stimulated different regions of the body with a 20 cm^2 thermal radiation exposure. They found that the face was most sensitive followed by chest, trunk and back followed by arms and legs. These results were similar to those found by Nadel et al. (1973) (see McIntyre, 1980).

Body maps show the distribution of physiological responses such as indicated by skin temperature and sweating as well as thermal sensitivity across the body surface. These have been derived for human subjects and much work has been conducted by George Havenith and his team in the climatic chambers at Loughborough University in the UK. Body maps for sweating and sensitivity have been produced. This follows on from the earlier work of Kuno (1956), where differences between Asian and western people were hypothesised.

Smith and Havenith (2011) constructed body maps for sweating for Caucasian male athletes during exercise in warm conditions. They found that greatest sweating was on the back and least on the fingers and thumbs. The forehead provided most sweating on the head. Gerrett et al. (2014) found that for a fixed warm stimulus (25 cm^2, 40°C) applied to 31 areas of the body, thermal sensation was greatest at the head, then the torso, hands and feet. Females were more sensitive than males and had a greater variation in sensitivity across the body. Sensitivity reduced during exercise, especially in males. Gerrett et al. (2015) conducted a similar experiment with female subjects and with the 25 cm^2 probe also at 20°C. For the cold stimulus the female subjects showed much greater sensitivity than for the hot stimulus. Most sensitive was the head followed by the torso then less in the extremities. A reduced sensitivity was again found during exercise. There seems to be an indication that sensitivity may be related to blood flow but this is not conclusive and requires further investigation.

The practical application of body maps may be in the design of clothing and environments. The relevance to local thermal discomfort is that it may be reasonable to assume that local thermal sensation and discomfort are related to the density, distribution and sensitivity of sensors and effectors over the body. One interesting finding by Stevens et al. (1974) is that the differences in sensitivity across the body reduced with the level of the stimulus applied. Hence when discomfort is considered the distribution of sensitivity may be different from that when a near threshold or even moderate stimulus is used.

LOCAL THERMAL DISCOMFORT CAUSED BY DRAUGHTS

ISO 7730 (2005) defines a draught as 'unwanted local cooling of the body caused by air movement'. This definition implies convective cooling, although it could include evaporative cooling if the skin is wet. Air movement also implies a small amount of pressure on the skin, although the term 'radiant draught' is sometimes used to describe the feeling of skin cooling by loss of thermal radiation to a cold surface such as a cold window. Hence no pressure on the skin caused by air movement so strictly outside of the definition. A draught may be caused from a cold surface, however, by cold air falling to the floor and causing discomfort to the feet and ankles.

Draughts can be felt across all parts of the body but are mainly detected in clothed people on the exposed skin of the ankles and neck, and maybe other areas depending upon clothing design. The cooling effect can cause discomfort but also pain and longer-term stiffness in muscles. In hot conditions, if local cooling is not extreme, air movement may be perceived as a welcome cooling breeze. When in neutral or cold conditions the same stimulus would be perceived as a draught.

Most studies of draught have blown air from tubes of various sizes and exposed skin (mainly to neck or ankles) in conditions where people are inactive and would have an (otherwise) overall neutral sensation. Parsons (2014) provides a fuller description with detailed references. ISO 7730 (2005) provides a Draft Rating (DR – originally termed Draft Risk but changed as risk implies danger in many languages). DR is the percentage of people dissatisfied due to draught and is given by DR = $(34 - t_a)(v - 0.05)^{0.62}(0.37v\,T_u + 3.14)$. DR is capped at 100% and t_a is the local air temperature (°C); v is the local mean air velocity (m s^{-1}) and T_u is the local air turbulence (%). This is sometimes termed turbulence intensity and is defined as the standard deviation of the local air velocity divided by the local mean air velocity expressed as a percentage. To feel a draught the air movement must interact with the air layer around the skin and reduce skin temperature. Turbulent air generally increases the disturbance of the air layer. As a 'time history', of air velocity is required to calculate its variance (standard deviation), when this is not possible, a value of $T_u = 40\%$ is often taken.

It is thought that for active people and people warmer than neutral (but not sweating), discomfort due to draught will be reduced for the same stimulus. It is possible that people who are cold overall will be more uncomfortable due to draught for the same local cooling but more research is needed to demonstrate this. A person who is cold will attempt to preserve heat so additional cooling will be unwelcome. This is consistent with the results of research into thermal pleasure (Cabanac, 1981), where providing a warm stimulus to a cold person or a cold stimulus to a warm person is pleasant, and a cold stimulus to a cold person and warm stimulus to a hot person is unpleasant.

If a person perceives a draught simultaneously in more than one area of the body, there is no guidance on what the overall DR would be. We can speculate from other modalities. It may be that we simply add the DR values or that we take the 'most severe component' (highest of the DR values) as the indicator of the overall effect. In acoustics a 'masking' approach is taken, where the most severe component reduces the effect of (masks) the other components so that they make a reduced contribution to the whole (greater than the highest DR but not an arithmetic sum).

An interesting exponential summation approach is taken for adding human vibration components. For example, the overall effect is reflected in the square root of the sum of the squares. If the DR is 10% for a draught at the feet and 20% for a draught at the head, the overall effect will be $(10^2 + 20^2)^{1/2} = 22.4\%$. The higher the exponent, the greater the weighting provided to the most severe component value.

Other methods include synergy where the total effect is greater than the sum of the parts. Research has not yet explored combinations of draughts or different types of local thermal discomfort effects as well as how to combine overall dissatisfaction with local dissatisfaction (e.g. Predicted Percentage of Dissatisfied (PPD) and Percentage of Dissatisfied (PD)).

When a person is hot but not sweating and air temperature is close to skin temperature, then there will be no discomfort due to draught. If the local moving air is cooler than the skin, then a local draught is possible but probably much reduced if the person is hot. For sweating and wet skin a draught is possible but there is little empirical research. A possible method is to conclude that the DR is related to convective cooling so it follows that any DR related to sweating and wet skin will be related to evaporative cooling. The Lewis relationship states that the ratio of the heat transfer coefficient by evaporation (he) to the heat transfer coefficient by convection (hc) is a constant (16.5 K/kPa). If the DR for heat transfer by convection is driven by the difference between comfortable skin temperature (34°C) and local air temperature (t_a) multiplied by hc, then heat transfer by evaporation is driven by the difference between the saturated vapour pressure at local skin temperature ($P_{s,sk}$ (kPa)) and the partial vapour pressure in the local air (P_a (kPa)) multiplied by he. As he = 16.5 hc, we can hypothesise that the percentage of people dissatisfied due to draught on sweating or wet skin is related to DR = 16.5 ($P_{s,sk} - P_a$)(v − 0.05)$^{0.62}$(0.37T_u + 3.14) capped at 100%.[1]

Although the DR value from ISO 7730 (2005) is derived from experimental studies on inactive males and females with an otherwise overall thermal sensation close to neutral, deviations from those conditions, as considered above, should be regarded as reasonable hypotheses yet to be tested (see Griefahn et al., 2001).

Two final points regarding draughts are that, in general, females are more sensitive to cold than males mainly due to smaller hands and fingers providing discomfort due to cold hands. It would therefore be expected that females would be more sensitive to draughts than males even when they wear identical clothing to that of males. As dress fashion in females often leaves legs and neck exposed (although leggings are often worn), a feeling of draught in females may be more prevalent. The final point for further discussion in the next chapter is that if a person is in a draught and the environmental conditions, individual ability, opportunity, disposition and context allow, people will move away from the draught. The draught can then be a stimulus for behavioural response rather than a sustained source of local thermal discomfort.

LOCAL THERMAL DISCOMFORT CAUSED BY RADIATION

Local Radiation

When predicting overall (whole body) thermal comfort, the mean radiant temperature (t_r) is used to represent the contribution of heat transfer by radiation to and from the body, that is the net average of all radiation into and out of the body in all directions. In some circumstances (particularly in solar radiation), radiation in some directions is more dominant than others. This is also influenced by a person's posture and hence 'projected surface area' towards

---

[1] As the relationship is empirical, a further multiplying constant may be required, but the principle should hold.

the directional radiation. Simon Hodder and I at Loughborough University in the UK, conducted a programme of laboratory and field experiments and found that for the standard thermal sensation scale (3, hot; 2, warm; 1, slightly warm; 0, neutral; −1, slightly cool; −2, cool; −3, cold) that, as an approximation, for every 200 W m^{-2} of direct solar radiation falling on a person, there was an average increase in sensation of 1 scale unit. For example, if a person is cool (−2) in the shade, and they are then exposed to the sun (e.g. 600 W m^{-2}) they will feel slightly warm (+1). That is 600/200 = 3 scale units from cool to slightly warm, −2 to +1). Although the research did not use focussed research on a small part of the body, in field experiments in vehicles, area of exposure varied. It was found, however, that the 'rule of thumb' held independently of surface area (and in laboratory experiments also independently of spectral content determined by different types of glazing) even though it is clear that the greater the surface area exposed, the greater the overall radiation input. For a fuller description, see Parsons (2014).

In addition to solar radiation, infra-red radiation to and from hot and cold surfaces provides radiation exchange and can cause local thermal discomfort. Solar radiation is of relatively short wavelength when compared with longer wavelength radiation for indoor and other lower temperature surfaces. This is because the wavelength of emitted radiation is shorter, the higher the surface temperature. Heat transfer by radiation between surfaces is proportional to the difference between the fourth powers of their absolute temperatures. If the skin temperature is higher than surface temperatures, there will be a net heat transfer out of the body; and if the skin temperature is lower than surrounding surface temperatures, there will be a net heat gain by radiation. This will have possible consequences for local thermal discomfort.

Radiant Asymmetry

Local thermal discomfort due to radiation is often caused by a single source from a single direction, for example, a furnace or boiler for people at work, direct or reflected solar radiation, machinery, heated walls, ceilings or floors, chilled ceilings, cold windows and more. A measure of radiation in one direction (the mean radiant temperature (t_r) integrates all directions) is the plane radiant temperature (tpr), and radiant temperature asymmetry (Δtpr) is the difference between the radiation from two opposite sides of a small plane element.

ISO 7730 (2005) provides the following percentage dissatisfied (PD) due to radiation asymmetry from a warm ceiling as PD = (100/(1 + exp(2.84 − 0.174 Δtpr))) − 5.5, for Δtpr < 23°C. From a cold wall, PD = (100/(1 + exp(6.61 − 0.345Δtpr))), and for a cool ceiling, PD = (100/(1 + exp(9.93 − 0.5Δtpr))) both for Δtpr < 15°C. For a warm wall, PD = (100/(1 + exp(3.72 − 0.052Δtpr))) − 3.5 for Δtpr < 35°C.

LOCAL THERMAL DISCOMFORT CAUSED BY VERTICAL TEMPERATURE DIFFERENCES

ISO 7730 (2005) provides the following equation for the percentage of dissatisfied due to temperature differences between head and feet (Δt_a,v) as PD = (100/(1 + exp(5.76 − 0.856Δt_a,v))), based on the research of Olesen et al. (1979). Although the experimental subjects were in thermal neutrality, it may be that any discomfort was caused by local air temperatures rather than air temperature differences. For experiments into comfort conditions for chilled ceiling and displacement ventilation environments, Loveday et al. (1998) found none or very little thermal discomfort over a wide range of temperature differences and humidity levels. Displacement ventilation is when cooled air (often accompanied by a chilled ceiling at a temperature above dew point to avoid condensation) is introduced to the floor by openings in the floor or from vertical 'bins' at floor level. People and machines (including computers) heat the air so that it rises and is 'exhausted' at the ceiling (so that it is breathed only once and any pollution is removed at the ceiling). In a field study in a modern office in the London city, displacement ventilation was used but people were complaining of headaches and drowsiness in the afternoon. On investigation, it was found that this was because of a build-up of carbon dioxide (from people) as workers at desks in the office had blocked the fresh air inlet ducts on the floor due to draught. Further research into vertical temperature differences is required as it is not clear if it is the difference that is causing discomfort or the local air temperature and air movement.

LOCAL THERMAL DISCOMFORT CAUSED BY WARM AND COLD FLOORS

When skin contacts with surfaces at skin temperature, there is a sensation of pressure. If the surface temperature of the material is at a higher temperature than the skin, then there will also be a feeling of warmth; and if it is at a lower temperature, there will also be a cool sensation. If the resulting skin temperature becomes very high or low, then it is possible that there will be pain and that the skin will be damaged (e.g. due to burning or freezing). Between extreme conditions (e.g. skin temperatures 5°C–40°C), there will be no damage but there may be discomfort. The magnitude of the (effect) sensation will depend upon the skin condition and the thermal conductivity (k), specific heat (c) and density (ρ) of the material of the surface, the nature of the surface, existing thermal sensation, the degree of contact and the contact time (see Parsons 2014, 2019). Practical examples of contact with surfaces of moderate temperature causing possible discomfort include cold floors, furniture including seats, handrails, tools, immersion in liquids, food types and clearly there are more.

It is a common observation that touching cold or warm metal will provide a greater magnitude of sensation than touching other materials such as wood or plastic, and this will occur even if they are at the same surface temperature. The reason is that the sensation is related to the rate of heat transfer between the material and the skin. A simple model, where there is perfect contact between two materials (with infinite capacity so that it can be assumed that neither material will change temperature), is to hypothesise a contact temperature (tc), which occurs directly on contact at the interface between the two surfaces. It is a weighted average of the material temperatures, the weighting determined by the thermal properties of the material. Contact temperature (tc) depends upon the thermal penetration coefficient of each material (b = $(k\rho c)^{1/2}$), so for two materials with values b_1 and b_2, tc = $(b_1 t_1 + b_2 t_2)/(b_1 + b_2)$. If b_1 is for the skin and b_2 is for the properties of the material surface in contact with the skin, then tc becomes the temperature at the skin surface. We can assume this to be surface skin temperature and if we determine the relationship between skin temperature and sensation, we can predict sensation on contact with any surface. Parsons (2014) provides detailed values as well as a method for estimating 'equivalent contact temperature' (tceq) that provides a method for taking account of contact type and skin condition.

Experiments with human subjects have been conducted on heated and cold floors as well as for handrails. Parsons (2014) presents data for handrails, where male and female young adults in neutral sensation reached into a thermal chamber containing handrails of different materials (wood, nylon, aluminium and steel) and rated their initial sensations and sensations after 20 s of firm but light contact normal for a handrail. The results were similar to those expected. Sensations at temperatures around initial skin temperature (neutral whole body sensation of 33°C) were rated as slightly warm and there was less than 10% dissatisfaction for all materials. At 5°C material temperature, aluminium and steel were rated as cold, with around 85% dissatisfaction. At 5°C material temperature, nylon and wood were rated as cool to cold and rated cool at 15°C with little dissatisfaction.

Fanger (1970) noted that complaints about cold or hot floors may be due to people being hot or cold, in general, as vasodilation and vasoconstriction will particularly affect the feet. He presents the work of Nevins et al. (1964a,b) for people in footwear. For bare feet, he calculates temperature ranges for the surface temperatures for different flooring materials from the equation for contact temperature (5°C–42°C for cork; 22°C–35°C for oak wood; 27°C–34°C for concrete; and more). McIntyre (1980) presents a review and more up-to-date temperature ranges for optimum conditions over a wide range of materials from 25°C for wood to 30°C for marble (see also Olesen, 1977, 1985 and Parsons, 2014). There are many countries particularly in Asia where under-floor heating and cooling is commonly used in homes and other buildings. As well as sitting on chairs and standing without shoes, sitting and lying on the floor are also natural postures. Although there has been some research into this important area, particularly with regard to energy reduction, more research is needed to investigate whether recommendations in current international standards apply to Asian people.

ISO TS 13732 Part 2 (2001) presents the relationship between human skin contact with materials at moderate temperature (10°C–40°C) and thermal comfort. It is noted that in a warm environment, cool surfaces may feel comfortable, and in a cool environment, warm surfaces may feel comfortable. Data are presented from experimental studies, a mathematical model of heat transfer (contact temperature) and from the use of a physical artificial foot. For 'neutral' environments comfort for hands, feet and sitting on the floor are considered. DIN 33 403-5 (1994) provides specification for an artificial foot and Marzetta (1974) for an artificial finger.

ISO TS 13732 Part 2 (2001) recommends that comfortable floor temperatures for people standing on typical floor constructions vary from around 20°C for carpet to 30°C for marble tiles. Dissatisfaction of people wearing normal shoes on heated and cooled floors is lowest at 25°C floor temperature at around 6% dissatisfied and at 15°C and 35°C floor temperature is around 20% dissatisfaction. No differences were found between men and women. For seated and standing people with shoes, it is recommended that floor temperature for floor heating systems should be lower than 29°C, floor insulation should maintain floor temperature above 19°C and that for warm to hot rooms floor temperature should be lower than 26°C. Special consideration is required for people who sit or lie on heated floors for long periods or have special medical or other requirements.

When people are exposed to local thermal discomfort the severity of effect will depend upon whether the person can adapt to reduce or eliminate the source, reduce its effects or move away from it altogether. Adaptive opportunity will be particularly relevant to avoid local thermal discomfort as well as to achieve thermal comfort overall.

Adaptive Thermal Comfort

6

ADAPTATION AND COMFORT

All living things from flowers and trees to reptiles, mammals and fish including and especially people continuously interact with their immediate surrounding environment and change their physical condition and behaviour in an attempt to achieve and maintain an optimum state. This ensures survival and in people is related to what we call 'comfort'.

Behaviour to achieve comfort is a universal human condition. It applies in all physical environments and is not restricted to certain types of environment such as types of building, vehicle or outdoors. It is a continuous process that is an integral part of the human disposition. To view it differently is a mistake and restricts progress towards an understanding of conditions that provide thermal comfort. Actually, it is not restricted to thermal environments, which are the consideration here. It applies to responses to acoustic, vibration, visual, air quality and other environmental components that combine to provide an integrated response to the total environment in which a person is immersed. People adapt to their environment, involving both physiological and behavioural responses, and in response to the thermal environment to achieve comfort this is termed 'adaptive thermal comfort'.

The use of behaviour to achieve survival has been recognised for many years, and in response to thermal conditions, it is termed behavioural thermoregulation. Jessen (2001) regards it as a means of reducing the demands on the autonomic system. He classifies mechanisms into locomotion (move to more favourable conditions); orientation and posture; wallowing (e.g. pigs in mud) and saliva spreading (over fur in rats); social behaviour (huddling) and operant behaviour (alter the environment). de Dear and Brager (1998) suggest that adaptation in the context of humans is the gradual diminution of the organisms' response to repeated environmental stimulation. This implies a learning process. They identify three types: behavioural adjustment (personal (e.g. alter clothing), technological (e.g. operate air conditioning), and cultural (e.g. take a siesta)); physiological (genetic and acclimatisation) and psychological (altered perception of, and reaction to, sensory information due to past experiences). They also note that the heat balance approach to predicting thermal comfort conditions, accepted in standards at the time, can be partially adaptive and does not conflict with but complements the adaptive approach to thermal comfort. There is no question that behavioural adjustment influences conditions required for thermal comfort, however it is not clear that physiological (as defined above) and psychological adaptation has influence in practice on thermal comfort requirements worldwide. It can be assumed therefore that adaptive thermal comfort is almost entirely related to behavioural response to thermal environments.

BIOLOGICAL ADAPTATION

Biological adaptation is well recognised in living organisms all of which exhibit both physiological and behavioural responses. Camels in the desert do not store water in their humps (although they can get water from fat at some expense to respiratory water vapour loss) but they do show adaptive responses to survive. For example, during dehydration (up to 30% of body weight) they vary their internal body temperature between high levels during the day and low levels at night. They also show rapid and accurate (not excess) rehydration within 15 minutes. The kangaroo rat, also in the desert, does not appear to drink any water at all and obtains its water from oxidation of, and any free water in, 'dry' food. Sheep, goats, cats and other animals use selective brain cooling when hot where cool blood from the 'wet' nose passes through a network (rete) of thin blood vessels in the carotid artery exchanging heat, and cooling blood, before it goes to the brain (Hardy, 1979).

Adaptation of animals to cold includes moving to a more favourable climate (e.g. seasonal migration of birds and animals and long-term drift as well as use of burrows), physiological adaptation to tolerate cold and hibernation. Physiological adaptation includes the use of anti-freeze (mainly glycerol), super-cooling and increased plasma insolubility to prevent freezing of cells, which is found in fish and some insects. Other adaptations include the increased ability for nerves to function in the cold in the limbs and feet of birds, increases in insulating fat layers and fur layer thickness in some animals and increased ability to produce metabolic heat involving brown fat and other mechanisms.

Behavioural thermoregulation can easily be seen in animals from birds by spreading wings to lose heat in the shade and gain heat in the sun, to my Persian house cat 'Nicky' who has a profound knowledge of where hot water pipes are in the house as well as where to lie in the sun coming through windows.

HUMAN ADAPTATION

Adaptation in people, as part of thermoregulation, was originally presented in the scientific literature as a set of peculiarities of people from 'distant lands' who live mainly outdoors. Examples include the Indians of Tierra Del Fuego (Land of Fire) at the southern tip of South America where lightly clad people carried fires with them to keep warm in the cold climate. Aborigine people in Australia and Lapps from Northern Finland were said to allow deep body temperature to fall at night, with reduced shivering while sleeping, to avoid sleep disturbance; and the female divers of Jeju Island of South Korea, and pearl divers from Japan, were said to inhibit shivering and enhance thermal insulation and vasoconstriction to reduce heat loss in cold water. People who fillet fish working in the cold were said to maintain hand blood flow even in cold water so that they could perform tasks without loss in manual dexterity performance (although there is a suggestion that this is caused by damage to hands). These changes along with acclimatisation, which enhances sweating after exposure to heat, are termed habituation and often gradually disappear after the exposure ceases. There are morphological (genetic) changes. People from cold climates have smaller noses and ears to avoid heat loss (as these areas have large surface area-to-mass (volume) ratio), and fat people have a greater thermal insulation –an advantage in cold climates.

The reader should consult the literature (e.g. Hardy, 1979; Jessen, 2001) for numerous examples of biological, including human, adaptation. How valid some of them are can be debatable. The point is that thermal physiologists and biologists tended to consider adaptive behaviour as specific phenomena. It is, however, a natural and inherent response to the environment. Imagine a (non-human) intelligent observer. They would immediately see that people are masters at adaptive behaviour and that this is because of their intelligence and ability to scenario test and predict and detect opportunities to adapt.

Whether a lobster can be comfortable or not is a debatable and maybe philosophical point that may come down to a matter of definition (our cat Nicky certainly seems to be comfortable in the sunshine). The term 'adaptive thermal comfort' seemed inappropriate when it was first used as adaptation had always been considered in terms of biological survival. Our independent (non-human) intelligent observer however would regard the 'naked ape' as a master of adaptation with a multitude of strategies from sweating in the heat to subtle and significant changes in clothing insulation to fans, air-conditioning, buildings and cities. Parsons (2019) describes human psycho-physiological thermoregulation as a form of homeostasis (holistic homeostasis), involving a continuous and harmonious interaction with environmental change. This is achieved by unconscious automatic thermoregulatory responses for survival combined with mainly conscious behavioural (adaptive) responses to maintain comfort with internal body temperature and thermal sensation as the controlled variables to be maintained at optimum levels (around 37°C and comfort).

Like all good paradigm shifts, adaptive thermal comfort complements rather than contradicts previous models. Debates of one versus the other are not useful. The Predicted Mean Vote (PMV) and Predicted Percentage of Dissatisfied (PPD) indices of Fanger (1970) as well as the Standard Effective Temperature (SET) index (Gagge et al., 1971) define comfort conditions based upon a model of physiological thermoregulation (internal body temperature, skin temperature and sweating). Adaptive thermal comfort adds to this paradigm to provide a more dynamic relationship between the environment and the achievement of thermal comfort based upon a psycho-physiological model of thermoregulation. In short, the achievement of thermal comfort is not a passive process as implied by, for example, the PMV index. Predicting the mean thermal sensation of a large group of people as 'cold' is a useful index of the group or even individual response to a static environment, but in terms of a human response and thermal comfort it should be seen not as a predicted final vote but as a drive for change. People do not wish to be 'cold' so will increase clothing insulation, move to a warmer environment or take whatever other perceived opportunities they have to achieve comfort. Non-adaptive thermal comfort approaches, such as the PMV and SET indices, have validity but they are limited. Rather than consider the PMV value as a thermal comfort index therefore it should be interpreted as the magnitude of drive for a behavioural response.

ADAPTIVE MODELS AND THERMAL COMFORT

Basing an understanding of how people achieve thermal comfort on the psycho-physiological model of thermoregulation directly enhances our ability and scope to ensure successful environmental design (actually not just for thermal comfort but by adopting the adaptive principle, for comfort, in general). Designing adaptive opportunities provides a range of physical conditions within which each individual can adapt to achieve thermal comfort. Not only will all people be comfortable and satisfied but the opportunities can be provided in an economic and energy efficient way (changing clothing usually has little cost and changing air conditioning 'thermostat set points' from 22°C to 25°C in hot conditions will save significant energy across a city). Parsons (2014) suggests that thermal comfort ranges can be provided in an adaptive way by the PMV index (increasing air velocity, changing clothing, etc. to achieve PMV = 0). However, this misses the dynamic nature of designing using adaptive opportunity.

It has always been known, from common experience, that people adapt (behave) to survive and achieve thermal comfort. However, the incorporation of adaptive approaches to thermal comfort applications only began around the early 1960s. One of the early pioneers was Auliciems (1981) who considered thermal comfort in offices and a psycho-physiological model of human perception. A landmark meeting of international researchers into thermal comfort took place at the Building Research Establishment (BRE) in the UK in September 1972. Rational (heat balance equation) approaches to predicting thermal comfort conditions and responses were presented by P O Fanger who presented his PMV/PPD indices. A P Gagge, Y Nishi and R Gonzalez presented the SET index and R G Nevins and P E McNall presented the newly proposed ASHRAE thermal comfort standard 55-72 as a basis for performance criteria for buildings. At the same meeting, J F Nicol and M A Humphreys presented thermal comfort as a self-regulatory system and in effect introduced the idea of adaptive thermal comfort (HMSO, 1973).

In the following years the rational (heat balance equation) approach, particularly the PMV comfort index, became accepted and indices were formalised into national and international standards used worldwide for setting criteria for indoor environments in buildings, for example (most recently ISO 7730 (2005), ASHRAE 55-2017 with addendum 2019 and more). Research into adaptive thermal comfort, mainly through field studies, however, continued with a summary of worldwide studies presented by Humphreys (1978) being particularly influential. He showed a relationship between the indoor temperatures measured in buildings and outdoor temperatures. So, for example, in hot climates, people seemed to prefer higher temperatures (for comfort). The implication is often taken that people in hot climates have adapted to prefer warmer conditions for comfort. This has not been demonstrated by research, which has confirmed early interpretations that this is caused by changes in clothing or air velocity, for example, and is not an inherent change in people. Conditions that dictate thermal comfort are in terms of the integrated effects of air temperature, radiant temperature, air velocity, humidity, activity level and clothing that people experience and can vary inside buildings independently of the outside climate. The potential for more versatile and successful environmental design, however, was clear, unfortunately progress has been slow for three reasons.

THREE REASONS FOR SLOW PROGRESS TO ADOPTING ADAPTIVE MODELS

The first reason was that the PMV/PPD 'method' had gained momentum and was controversial. Much research was conducted into their empirical validity as well as into the underlying assumptions made in their development. Is 'thermal load' really a good indicator of thermal sensation and dissatisfaction for conditions away from neutrality? Are the PMV/PPD indices applicable for all populations across the world? Do people from Japan require the same conditions for thermal comfort as people from the USA? The questions were important and there is no doubt that the PMV/PPD indices were a significant improvement on what went before. The focus, however, was away from, and restricted progress in, the area of adaptive thermal comfort.

The second reason for slow progress towards adaptive approaches was the lack of clarity about what adaptation meant in practice and what was actually occurring. The implication was that this was behavioural thermoregulation, particularly changing clothing and activity level, but also opening windows, moving around and many other behaviours, often linked to population culture. Parsons (2014) showed that these behaviours alone would change the integrated effects of air temperature, radiant temperature, air velocity, humidity, clothing insulation and activity level such that the PMV index would be a valid predictor of preferred air temperature. Adaptation, however, was also attributed to habituation and changing physiological and psychological state. Do people from hot climates sweat more and hence accept more sweating for thermal comfort conditions? Have people in cold climates learned to restrict vasoconstriction to maintain comfort? If so, the actual reason for a variation in comfort conditions would not solely be behavioural but also physiological adaptation. Psychological changes have also been suggested. There is some evidence that people do not complain about being 'too hot' because they do not expect any better (Fanger and Toftum, 2002). This is a complicated point. Should we determine thermal comfort conditions because we can 'get away with it'? People may still be dissatisfied but not inclined to complain and in any case there is no evidence to suggest that people who do not complain at 30°C air temperature even if comfortable, for example, would complain about being too cold or be dissatisfied at 24°C (e.g. PMV = 0). It has also been suggested that if people have control of their environment then they will accept (be comfortable in) a wider range of conditions. At least they will be able to match the environment to their individual requirements so maybe zero dissatisfaction with sufficient adaptive opportunity through environmental design is optimum. People won't complain as they can do something about it. Actually, this is not different from the principle of behavioural thermoregulation and adaptive thermal comfort.

A third reason for slow progress towards the adoption of the adaptive thermal comfort paradigm is that it was taken over by researchers with a building-centred approach when the adaptive paradigm is essentially a human-centred phenomenon. There was a focus on the 'type of environment (building)' rather than recognising that adaptive comfort is a human disposition, which people bring to all environments. A human-centred approach will consider comfort as the controlled variable in a continuous psycho-physiological process of seeking and preserving thermal comfort. The building-centred approach led to suggestions that adaptive comfort was specific to types of building (applicable in naturally (free running) ventilated buildings, much influenced by outside conditions, but not in air-conditioned buildings generally isolated from outside conditions with sealed windows, etc.). The problem was then confounded, and progress hindered, by the underlying assumption that the causal model in buildings for adaptive thermal comfort was a weighted sum of variables (exponents of variables, interactive combinations, modifications for time lags, etc.). That is correlation and regression techniques. This was useful for building engineers as they use outside climate data to determine how to create appropriate indoor environments. However, it is reasonable to assume that people will respond to the environments that they perceive (indoors), and that there is no fundamental scientific reason that this should be a regression model. The best we can get with a regression model is that if something correlates well then it may have sufficient predictive power for a particular application. There is little dispute that any causal model of human response to a thermal environment must include air temperature, radiant temperature, air velocity, humidity, clothing and activity level. Adaptive behaviour must complement that and any attempt to use a simple regression model will provide a solution of low general utility. That is not to say regression models did not provide a starting point and when Humphreys (1978) suggested a regression model involving a prediction of preferred indoor temperatures in free running buildings from mean outside temperatures it opened up the debate and allowed movement towards an adaptive approach to thermal comfort.

There has always been an ambiguity in adaptive interpretations of field data, of whether people change or populations are different so that they require different comfort conditions, or if people behave (maybe in different ways depending upon culture) in response to environments to achieve and preserve comfort. The early notion of thermal comfort as a self-regulating system involving autonomic and behavioural thermoregulation (Nicol and Humphreys, 1972) to the Adaptive Principle "If a change occurs such as to produce discomfort, people react in ways that tend to restore their comfort" (Nicol et al., 2012) suggests behaviour change but the notion that different populations of people have different requirements continues to be proposed.

ADAPTIVE THERMAL COMFORT REGRESSION MODELS

The advantage to building engineers of predicting comfort temperatures from outside conditions is that the weather data are available for locations throughout the world. Humphreys (1978) demonstrated that preferred temperatures for buildings (particularly free running buildings) is linearly related to outside temperature. In a recent interpretation, Humphreys et al. (2010) proposes that the comfort temperature $T_{comf} = 0.53(T_{om}) + 13.8°C$, where T_{om} is the mean of the monthly mean of the daily maximum outdoor temperature and the monthly mean of the daily minimum outdoor temperature. There are other equations for different conditions (for a summary, see Nicol et al., 2012).

The American Society for Heating, Refrigerating and Air-Conditioning Engineers (ASHRAE, 1998) provided one of the first formal attempts to introduce the adaptive thermal comfort paradigm into thermal comfort standards. A landmark workshop was held in San Francisco in 1998 to present worldwide field studies and bring together data and ideas on thermal comfort and adaptation. A global database of thermal comfort field experiments and how this was used to develop an adaptive model were described by Richard De Dear and Gail Brager who, with others, had conducted the extensive worldwide survey on behalf of ASHRAE (the ASHRAE database). There was some difficulty over selecting which outside environmental conditions should be used to predict indoor comfort temperatures. The use of air temperatures alone omitted important solar radiation and other components. An attempt to use New Effective Temperature (ET*) provided more validity but weather data are not available for all of the variables in the form required for ET* so assumptions had to be made. The standard ASHRAE 55-2004 provided the optimum temperature for comfort as $T_{comf} = 0.31T_o + 17.8°C$, where T_o is loosely defined as the prevailing mean outdoor temperature. European standard, EN 15251 (2007) used data from European studies to provide the adaptive model $T_{comf} = 0.33T_{rm} + 18.8°C$, where T_{rm} is the exponentially weighted running mean of the outdoor temperature (see Nicol et al., 2012).

The international standard ISO 7730 (2005), mainly concerned with the PMV/PPD indices, comments on the adaptive approach but does not present an adaptive model as it is a human-centred standard not specifically concerned with buildings. There is a proposal to develop an adaptive thermal comfort standard based upon the adaptive approach and adaptive opportunities but this has yet to get off the ground.

ADAPTIVE MODELS THAT MODIFY THE PMV INDEX

An adaptive approach could be taken by modifying the PMV index. Fanger and Toftum (2002) suggested that if people who did not expect any better would not be dissatisfied with or complain about a thermal environment, then the PMV value could be multiplied by an expectancy factor, reducing the PMV severity and hence the PPD. However, that people would expect no better would seem to be an inappropriate method of determining and specifying conditions for thermal comfort.

Parsons (2003) noted that for most environments the most frequently used, and effective method of, adaptation was to alter clothing insulation. He suggested that all adaptive opportunities could be integrated into an equivalent adjustment to clothing value (I_{EQUIV}) that would give an effective clothing insulation to be used in the PMV equation, hence providing a PMV value taking account of adaptive behaviour. The formulation of the method limits the effective clothing insulation from 0 to double the initial clothing before adjustment (I_{START}). While this is a pragmatic proposal applicable over many applications, this restriction makes the method cumbersome. The equivalent clothing insulation is defined as "...the clothing insulation that would give equivalent thermal comfort to people with no adaptation as the thermal comfort of people who adapt to their environment". It is identified that adaptation to discomfort in heat and cold can have different mechanisms. So $I_{EQUIV} = I_{START} - (I_{ADJ} \times I_{START})$ for 'warm discomfort' in heat and $I_{EQUIV} = I_{START} + (I_{ADJ} \times I_{START})$ for cold discomfort, where I_{ADJ} has a value of adaptive opportunity of 0, minimum; 0.25 low; 0.5, medium; 0.75, high; 1, maximum. So there is a reduction of clothing in the heat and increase of clothing in the cold.

Yao et al. (2009) proposed an adaptive PMV index aPMV = PMV/$(1 + \lambda PMV)$, where the PMV is the Predicted Mean Vote determined for environmental, clothing and activity conditions and λ is the adaptive coefficient which is empirically determined for different contexts. For data in Chongqing, China, for example, λ was found to be 0.293 for warm conditions and -0.125

for cold conditions. For this method, the values of λ need to be determined for the contexts of interest, and it is not certain which particular adaptive opportunities are available or have been taken. There is also a suggestion of psychological adaptation and the method also considers the PMV as a whole rather than considering the influence of the adaptive behaviour on its component parts.

A more causal and hence powerful approach to determining the effects of adaptive behaviour is to consider the effects of the adaptation on the conditions to which a person or people are exposed. For example, adjusting clothing influences clothing insulation, slowing down work rate reduces metabolic heat production, turning on fans increases air velocity and so on. This would provide a more reliable predictive method than general methods described above. Oseland et al. (1998) analyse different methods of adaptation and suggest temperature offsets for neutral temperatures due to different adaptive behaviours (e.g. operate a desk fan at $2\,\mathrm{m\ s^{-1}}$, increases neutral temperature by $2.8°C$). As each adaptive opportunity is often a complex interaction with the thermal conditions experienced over time, however, this is desirable but difficult to achieve in practice.

Parsons (2019) proposed a psycho-physiological system of human thermoregulation that includes adaptation to avoid heat stress. This is presented in modified form in Figure 6.1. It shows the autonomic response to thermal conditions controlled by the hypothalamus, to maintain internal body temperature, in combination with the conscious adaptive response (to additionally attain and maintain thermal comfort).

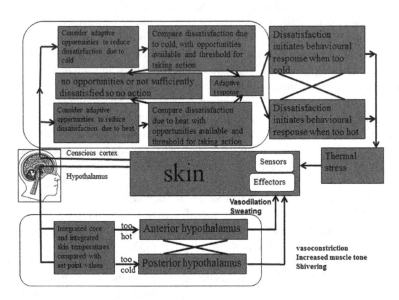

FIGURE 6.1 Psycho-physiological system of thermoregulation.

For a comprehensive review and description of adaptive thermal comfort, the reader is referred to a trilogy of books by Nicol et al. (2012), Humphreys et al. (2014) and Roaf et al. (2014).

For progression in our understanding of, and design for, thermal comfort, it is important to build upon what has been achieved but also to leave behind methods that do not include adaptive thermal comfort. The interaction of the seven factors that make up environments for thermal comfort (air temperature, radiant temperature, humidity, air velocity, clothing, activity and adaptive opportunity) must be considered together to specify comfort conditions. A representation of how an adaptive approach can be shown on a psychrometric chart is provided in Figure 6.2.

Comfortable environments will be those that allow comfort to be achieved by each individual using adaptive opportunities and behaviours recognised in environmental design. If this is achieved, it will open the door for not only comfortable environments across the world but those that are energy efficient, delightful, inspirational and give pleasure.

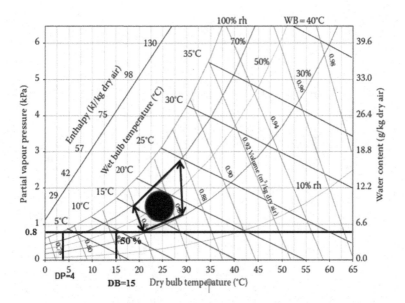

FIGURE 6.2 Psychrometric chart showing aspirated wet bulb temperature at 10°C, dry bulb = 15°C so rh = 50%, Dew Point (DP) = 4°C, Pa = 0.8 kPa, humidity ratio = 6 g kg⁻¹ of dry air. Black dot shows region of PMV = 0 for 50% rh and 0.8 clo. Area of acceptability for adaptive comfort is bounded by recommended rh of 30%–70% and line for PMV = –1 (slightly cool) and 1.0 clo to PMV = +1 (slightly warm) for 0.6 clo. So this is a region of adaptive comfort where clothing can be adjusted between light and medium insulation.

Thermal Comfort in Planes, Trains, Automobiles, Outdoors, in Space and under Pressure

7

SPECIAL ENVIRONMENTS

The previous chapters, particularly Chapter 6 on adaptive thermal comfort, have made the case that there are no special environments as the human disposition is to pursue survival and thermal comfort and the principles for achieving this do not vary with the type of environment. There are environments, however, that differ from what could be regarded as the 'normal' steady-state indoor office environment, particularly when exposure to the thermal conditions varies with time and space. These include vehicle environments where people often occupy relatively small spaces with limited adaptive opportunity and are exposed to the changing external environment through windows and to significant contact with surrounding surfaces, as well as other environments such as those in space, hypo- and hyper-baric environments and outdoor environments.

THERMAL COMFORT IN AUTOMOBILES

Most studies of thermal comfort have been related to buildings, however there have been some related to vehicles and particularly automobiles including motor cars. This has been mainly with gasoline-driven vehicles, where sufficient energy is available for heating, ventilation and air conditioning, albeit at a cost. A challenge for the future is to provide thermal comfort in 'green' vehicles driven, for example, by electricity where limitations in power provide a requirement for innovative ideas. The principles of designing for thermal comfort will not change. People will respond to the same parameters in similar ways. The unique requirements of some environments may, however, require special consideration. Although the adaptive thermal comfort paradigm has now become established, this has yet to penetrate studies in vehicles and more complex environments, where in fact it may be particularly relevant. In addition, vehicle environments in particular are influenced by direct solar radiation through windows as well as rapid changes in conditions. The movement of a vehicle will ensure that the outside surface of the vehicle will achieve the temperature of the air. Glazing, such as used in glass cabs to provide good visibility, can, however, cause thermal discomfort from the sun. When the vehicle is stationary, solar radiation, in particular, will heat the vehicle surface, and together with direct solar radiation through windows can heat vehicle interiors to high and intolerable levels.

O'Neill et al. (1985) provided one of the first modern studies into the thermal comfort of tractor drivers. This led to a demonstration that a canopy to provide shade from the sun would be useful and eventually led to darkened glazing and air-conditioned cabs becoming the 'norm'. Rohles and Wallis (1979) concluded that, for thermal comfort, cool air should be blown towards the chest of drivers and passengers in cars. Wyon et al. (1985) used a heated thermal manikin to investigate thermal comfort in trucks and cars as well as introducing ventilated seats. They suggested a thermal comfort index for vehicles termed the Equivalent Temperature. This involved the distribution of temperatures across a person and was adopted by the ISO 14505 series of international standards (see also Madsen and Olesen, 1986).

Equivalent Temperature was defined as the uniform temperature of an imaginary enclosure with still air, in which a person would exchange the same dry heat, by radiation and convection, as in the actual environment. It therefore uses the standard environment approach to define equivalence, although the index value has yet to be directly related to thermal comfort. By definition, it does not include sweating and heat loss by evaporation, which are often causes of discomfort in vehicles. It is particularly relevant to the evaluation of asymmetric environments and measurement typically involves the use of thermal manikins although other methods are acceptable. The Equivalent Temperature index was extensively studied and developed by a multi-national European research project (EQUIV), which greatly influenced international standards ISO 14505 Parts 1–4 (Nilsson et al., 1999). The Society for Automotive Engineers in fact had adopted the Equivalent Temperature index from earlier research (SAE, 1993).

Hodder and Parsons (2001a,b) were asked, by glazing and car manufacturers, to determine the relationship between solar radiation and thermal comfort in a multi-national European project to determine the effects of glazing. Glass provides a significant contribution to overall vehicle weight and hence fuel consumption. Reducing the weight of glass may, however, change the thermal transmission properties of the glass and hence influence thermal comfort. An understanding of the effects of solar radiation on thermal comfort, particularly through glass, was required. A thermal simulator was constructed (Hodder, 2002) to investigate the effects of (simulated) solar radiation of different orientations, levels and spectral content, on the thermal comfort of seated passengers. A standard protocol was adopted for the simulator experiments where participants dressed in light clothing, sat in a car seat and were exposed to simulated solar radiation through glass for 30 minutes of exposure. Ratings of thermal sensation (−1, slightly cool; 0, neutral; 1, slightly warm; 2, warm; 3, hot; 4, very hot; 5, extremely hot), discomfort, stickiness, pleasantness and acceptability were taken every 5 minutes throughout the exposure. The results consistently showed that there was a linear increase in thermal sensation vote with level of direct solar radiation. Over the range 0–600 W m^{-2} direct solar radiation on the person, for every increase of around 200 W m^{-2} the average thermal sensation increased by 1 scale unit. It was found that sensation was related to level (intensity) of solar radiation and not to spectral quality (as determined by glazing type) and that the 'rule of thumb' of 200 W m^{-2} for a rise of 1 scale unit, applied across the thermal sensation scale. For example, if the thermal conditions were 'neutral' (e.g. Predicted Mean Vote, PMV = 0), adding 400 W m^{-2} direct solar radiation (moderate sunny day) would increase average sensation to +2, warm. If the thermal conditions were −2, cool, adding 600 W m^{-2} direct solar radiation (sunny day) would increase average sensation to +1, slightly warm and so on. It was also found that re-radiation from the dashboard had relatively little effect with up to 0.5 scale units increase for a black dashboard with a surface temperature of 100°C.

The results from the laboratory simulator studies were evaluated in extensive field trials over 10 consecutive days in the summer in Spain. Subjective ratings were taken from male passengers driven (in Range Rover Freelander cars where glazing was changed overnight) from Seville to Cadiz and back (most consistent sunny blue sky in Europe and often used for vehicle manufacturer tests). A comparison of a range of methods for predicting thermal sensation votes showed a high correlation with other subjective measures and lower but (statistically) significant correlations for other methods. Of note are the simple method of using the temperature of the dashboard as a prediction method and a method of using the PMV index in the shade with an addition of the effects of direct solar radiation on the person. Although far from perfect, this gave encouragement to the validity of the laboratory experiments and as a rational thermal index, such as the PMV index, includes variables of known importance to thermal comfort (air temperature, radiant temperature, humidity, air velocity, clothing, activity) an index of thermal comfort for conditions in direct solar radiation was proposed. The proposal was $PMV_{solar} = PMV + RAD/200$, where PMV_{solar} is the Predicted Mean Vote of people in direct solar radiation (−5 extremely cold; −4, very cold; −3, cold; −2, cool; −1, slightly cool; 0, neutral; 1, slightly warm; 2, warm; 3, hot; 4, very hot; 5, extremely hot), PMV is the Predicted Mean vote (ISO 7730, 2005) taking account of insulation provided by the seat (e.g. add 0.2 clo) and for mean radiant temperature in the shade, and RAD is the direct solar radiation on the person (1,000 W m⁻², maximum; 600 W m⁻², sunny; 200 W m⁻², cloudy with little sun; 0 W m⁻², minimum) Parsons (2014). This model required further validation, however, it provides some validity from empirical studies and includes the accepted relevant variables in a logical way.

In an attempt to provide a more personal (adaptive) approach for thermal comfort in motor cars (also energy efficient), Brooks and Parsons (1999) investigated the use of a 'new material' that could effectively heat seats in a uniform and easily adjustable way. (Actually the material is currently used to heat racing car tyres in Grand Prix racing.) They found that people could maintain comfort over a surprising range of temperatures with a limiting factor for the driver being cool hands on the unheated steering wheel, and eventually cold feet for all. An advantage of this method is that each individual in the vehicle can control their own seat temperature. The thermal properties of vehicle seats and their influence on thermal comfort have been extensively reported by Fung and Parsons (1996) who conducted a series of climatic chamber experiments with human subjects.

THERMAL COMFORT ON TRAIN JOURNEYS

A journey on public transport requires leaving home or other base, travelling to a 'station', waiting often outdoors, getting on the vehicle (bus, train, etc.), riding on the vehicle, getting off the vehicle and then travelling to a final destination. This usually requires the traveller to be exposed to a range of thermal environments. Kelly (2011) considered the whole of a train journey. The traveller may arrive at the station in a number of thermal states. As there are numerous methods for getting to and from a railway station, we will concentrate on thermal comfort from while waiting for the train, through the journey, to alighting from the train.

The laboratory simulator studies described above for thermal radiation and comfort will apply to trains as well as to motor cars. Additional studies, using a similar protocol, were carried out to enhance information relevant to trains. Experiments with clothing fit and colour showed that tight fitting black clothing was rated hotter than loose fitting white clothing by about 1 scale unit for the same thermal conditions (Parsons et al., 2005). Similar results to those of Hodder and Parsons (2001a) were found by Vaughan et al. (2005) for both front-on and side-on radiation. When participants sat next to a cool window (6°C), Underwood and Parsons (2005) found reduced sensation (cooler) on the side adjacent to the window of around 1 scale unit. A draught was also found indicating that a cool window causes cool air to fall and move across the ankles as well as an asymmetric radiation 'draught'. Stennings (2007) investigated exposure time and found that there was little difference between the experimental protocol exposure of 30 minutes and that of 1 hour. After the 60 minute exposure, a screen was used to block any simulated solar radiation arriving on the participant and then after a further 5 minutes the screen was removed (e.g. as for the train going into and out of a tunnel). It was found that for both conditions, participants immediately experienced sensations that they would have experienced in steady-state conditions. Kelly (2011) conducted two laboratory experiments in thermal chambers using a total of 48 male and 48 female participants to investigate conditions when getting on and off trains. It was found that, when compared with expected sensations from the (steady state) PMV index, there were no overshoots or undershoots (cooler or warmer than expected) when people went from cooler to warmer environments but that overshoots did occur when going from warmer to cooler environments. Kelly (2011) investigated thermal comfort while getting on and off trains and found overshoots in both directions (cool to warm and warm to cool). The role of air velocity may be influential in field studies where moving air is not uncommon on railway station platforms and air movement is effective at removing the insulating air layer on bare skin (as well as creating pressure). Moving from the station platform with significant air movement, to the train with still air, or vice-versa, will cause a thermal shock and maybe some 'over-reaction' due to a high rate of change of skin temperature. Investigations on trains (using branch lines as well as high-speed trains between Loughborough and London) by Underwood (2006), Stennings (2007) and Kelly (2011) demonstrated the usefulness of the PMV_{solar} method for assessing thermal comfort in vehicle environments.

Crowding is a major factor in commuter and other train journeys. In addition to psychological stress, other passengers will add to radiation input and convective and evaporative heat transfer will be restricted (useful in cold climates but uncomfortable in warm to hot conditions). Braun and Parsons (1991) developed a crowding correction factor based on results of a laboratory study on physiological and subjective responses to thermal environments and crowd density. Parsons and Mahudin (2004) extended this work to develop a risk assessment method for assessing crowds that included measurements in the 'rush hour' on the London Underground.

THERMAL COMFORT IN AEROPLANES

Interior environments for passengers on aeroplanes are entirely controlled due to the hostile external environment. There is little contribution of direct solar radiation through windows to passengers, and although the passengers have limited adaptive opportunity, they can move around to some extent and can usually adjust air velocity (jets) from above. As we go higher into the atmosphere, atmospheric pressure reduces providing less oxygen for people to breath. In passenger aeroplanes, this is counteracted by 'pressurising' the cabin to the equivalent of being at around 1,800 m (around 6,000 feet) above sea level. In this hypo-baric environment, heat transfer by convection is reduced compared with the same thermal conditions at sea level and heat transfer by evaporation is increased. Parsons (2014) provides simple modifications to equations for heat transfer to and from the body by respiratory, convective and evaporative heat transfer due to breathing and by heat transfer from the skin by convection and evaporation. If PB is the atmospheric pressure in atmospheres, then convective heat transfer due to breathing is Cres = Cres at sea level × PB; for evaporation, Eres = Eres at sea level/PB; the air velocity term (v) in the heat transfer by convection equation becomes v × PB; and the Lewis relation becomes LR/PB so the evaporative heat transfer coefficient he = 16.7 hc/PB $m^{-2}kP_a W^{-1}$ (see Chapters 2 and 4 and Parsons, 2014).

Nishi and Gagge (1977) present comfort temperatures for hypo- and hyper-baric environments based on their calculation of their New Effective Temperature index (ET*) with modified heat transfer equations. For non-sweating, lightly clothed people in still air, they found that there were only small differences (0.2°C) between those required in pressurised aircraft cabins and those at sea level. This is insignificant relative to any effects of adaptation, personal control, draughts caused by air movement (Liu et al., 2013) and the thermal properties of the seat as well as the influence of other passengers. Although modifications can be made in a similar way to the PMV thermal comfort index equation to account for cabin pressure, the effects are small. It is probably reasonable in design to begin with the existing PMV index (ISO 7730, 2005), consider local thermal discomfort (maybe modified for the hypobaric environment) as well as designing in adaptive opportunities, as the main drive for design of comfortable thermal environments in aircraft cabins.

THERMAL COMFORT IN SPACE VEHICLES

One of the main differences between aircraft environments or buildings on earth and those in space is the near complete absence of gravity. There will therefore be no heat transfer by natural convection (and linked influence on evaporation) in space. Warm, less dense air, heated by the body, will not rise and cooler heavier air will not fall due to gravity and replace it. For space stations, buildings and structures on other planets, this will depend upon the local level of gravity present. Parsons (2014) suggests that in space, air temperature, radiant temperature, humidity, air velocity, clothing insulation and activity level as well as air pressure will all influence thermal comfort. Heat transfer equations and hence the human heat balance equation will apply (in modified form from that on earth). Heat transfer mechanisms by convection and evaporation may differ, relative air velocity may become significant, sweat distribution will change and metabolic rates will be reduced as the body will not have to overcome gravity. With small modifications to the heat balance equation, the ET* and PMV rational indices will provide a starting point for comfort temperature ranges. Much further research is needed however. It is not known how physiological responses will relate to comfort conditions. What sweat rates or skin wettedness levels will be appropriate for comfort? Will skin temperatures vary as on earth and how will these relate to comfort, discomfort and dissatisfaction and so on. In the future, it will be necessary to conduct empirical research with human subjects to determine if existing methods for specifying thermal comfort conditions apply (in modified form) or whether new methods and approaches are needed. Adaptive approaches to thermal comfort will be essential and, if adopted from the initial design requirements for the environment, will lead to innovative solutions.

OUTDOOR THERMAL COMFORT

Outdoor thermal environments are determined by weather conditions and thermal comfort will be particularly influenced by solar radiation and relatively high and volatile air velocities. There is great interest in providing pleasant, stimulating and comfortable outdoor environments in cities, for example. There is a category of outdoor environments that people seek for pleasure, such as when sunbathing on a beach. These are 'special' environments that should be treated separately from 'ordinary' thermal comfort conditions.

Thermal pleasure and delight are of great interest to architects and environmental designers however, and further research into creating such environments is required. de Dear and Spagnolo (2005) considered outdoor environments where they are an extension of indoors and those where they have specific characteristics due to being outdoors. It was proposed that thermal comfort in the former can be determined using accepted indoor environmental indices such as the Standard Effective Temperature Index (SET). They used this to calculate required air velocity for comfort for people in Disneyland queues in the USA.

Adaptive approaches are particularly relevant to provide comfort in outdoor environments. Seeking shade, selecting and adjusting clothing, finding a cool breeze or a sunny alcove out of the wind are all examples of how people maintain thermal comfort outdoors.

The clothing required (IREQ) for comfort was part of an ISO standard mainly concerned with cold stress (ISO 11079, 2007). People select clothing for many reasons however, particularly for fashion and self-image, and they plan their journeys outdoors and indoors in an attempt to maintain thermal comfort. The development of the SET by Gagge et al. (1971) was an attempt to provide an index that applies to outdoor conditions.

The Universal Thermal Climate Index (UTCI) was developed by a multinational team of researchers to predict human responses, including thermal comfort, to outdoor conditions (Jendrinsky and de Dear, 2012). The objective was to link predictions of human response with weather data and predicts thermal comfort conditions as a rating of 'no thermal stress' (UTCI = 19.3°C). Based upon a mathematical model of human thermoregulation (Fiala et al., 2012) and an integration of up-to-date information on human response to thermal environments, a method was determined to calculate the UTCI for any outside environment. The UTCI is defined as the air temperature of a reference environment that would give equivalent thermal strain (human response) to that of the actual environment (Bröde et al., 2012). The reference environment is 50% relative humidity (capped at $20\,hPa = 2\,kPa$), calm air ($0.5\,m\,s^{-1}$

at 10 m above the ground or 0.3 m s⁻¹ at 1.1 m) and radiant temperature equal to air temperature. The activity level is assumed to be walking at 4 km hour⁻¹ (135 W m⁻²), and the standard clothing was calculated using a formula involving air temperature. The actual environment is made up of any combination of air temperature, radiant temperature, humidity and wind speed, and equivalence is defined as a combination of seven predicted physiological responses (rectal temperature, mean skin temperature, facial skin temperature, sweat production, skin wettedness, skin blood flow and shivering).

At the time of publication a UTCI calculator can be found at www.utci.org/utcineu/utcineu.php. For an air temperature of 20°C, mean radiant temperature 10°C greater than air temperature, 50% relative humidity and wind speed 3.0 m s⁻¹ at a height of 10 m above the ground, the UTCI is calculated at 19.3°C implying no thermal stress. The UTCI is intended to be a universal index based upon the weather and ranging from hot to cold environments. For thermal comfort outdoors, it will provide a first approximation to human responses and required conditions.

THERMAL COMFORT IN HYPER-BARIC ENVIRONMENTS

It is unusual to consider discomfort in hyper-baric environments. As for hypo-baric environments, any small deviations in atmospheric or other air pressure will have only small effects on thermal comfort. Fanger (1970) suggests meteorological changes in pressure will have insignificant effects but suggests that deep mines, pressurised chambers for medical care and ocean floor sea laboratories are all special hyper-baric environments. We can add to this hyper-baric lifeboats, where divers await rescue if a 'mother-ship' becomes unavailable and compressed air tunnels that are used to keep water from nearby sea, lakes or rivers at bay. Nishi and Gagge (1977) and Parsons (2014) provide a discussion and analysis. Opposite to low pressure, high pressure increases heat transfer by convection and reduces heat transfer by evaporation. The modifications to equations provided above for hypo-baric environments will apply, (PB (atmospheres) will now be greater than 1). In very high-pressure environments, ranges of comfort temperatures become narrow, and careful control is needed to avoid extreme environments and rapid decline into dissatisfaction. O'Brien et al. (1997) reported a study of workers in sealed hyper-baric tunnels for the new Jubilee Line Extension for the London Underground. Pressure was required to prevent the river Thames from entering the works. Heat stress was a major problem, and although discomfort was not the major concern, it can be speculated that the lack of ability to evaporate sweat led to stickiness and dripping due to wet skin. More fundamental and systematic studies into discomfort in hypo- and hyper-baric environments as well as in space and on other planets will be required if people in the future are to inhabit and occupy dwellings in what would presently be regarded as non-typical or hostile environments.

THERMAL COMFORT UP MOUNTAINS

Up mountains people experience lower atmospheric pressures than they would experience at sea level as there is less atmosphere on top of them. Air temperature, radiant temperature, air velocity, humidity, clothing and activity remain the important factors however heat transfer between a person and the environment is affected by pressure, reducing convection and increasing evaporation. Solar radiation levels increase greatly up mountains, air temperatures typically fall, wind increases and we often experience clouds, rain, hail and snow. Tents have low thermal inertia so are greatly influenced by outside temperatures. There will be a decrease in oxygen level available to the body in hypobaric environments which can lead to discomfort (and much more).

Parsons (2014) suggests that heat transfer by respiratory convection (Cres) is Cres × PB (Wm^{-2}) where PB is the atmospheric pressure in atmospheres (1.0 at sea level, less up mountains, more in pressurised air). Heat transfer by forced convection modifies the heat transfer coefficient by convection (hc) to hc = 8.7 (v × PB)$^{0.67}$ (v in ms^{-1}, with other adjustments for natural convection) and for evaporative heat loss uses he = 16.7 hc/PB. Fanger (1970) concludes that daily meteorological variations in pressure up to 3000m above sea level, do not require a modification to his comfort equation. As pressure decreases, heat transfer by convection would decrease and evaporation would increase (enhancing dryness). Nishi and Gagge (1977) suggest that for low atmospheric pressures (e.g. PB = 0.33) such as those found up mountains, such as Mount Everest, comfort temperatures were predicted to be similar to those at sea level, and comfort and discomfort related to mean body temperature. An important lack of knowledge of thermal comfort up mountains, is in the relationship between environmental conditions, physiological state and degree of thermal discomfort.

Thermal Comfort and Gender, Age, Geographical Location and for People with Disabilities

8

THERMAL COMFORT FOR ALL

It is remarkable that to a first approximation, thermal comfort conditions are similar for all people if the air temperature, radiant temperature, humidity and air velocity to which they are exposed as well as their clothing and activity are taken into account and we include their mechanisms and capacity for adaptation. As essentially the human thermoregulatory system is similar for all people then for healthy people there will be similar physiological responses to heat and cold. However, each individual person in the world has unique characteristics so not surprisingly there are individual differences in responses to thermal environments.

Current considerations of thermal comfort, including variation in response, are based upon the assumption of homogeneous populations. Olesen (1982) notes that in a study of 64 subjects the standard deviation on the preferred ambient temperature was 1.2°C. Interestingly McIntyre (1980) reports on intra- (changes within the same subject with time) as well as inter- (differences between the responses of different subjects) subject variation and found them to be similar. This implies that predictions for average responses can be used for predictions of individual responses with similar validity although there may be other interpretations and more studies of intra-subject variation are needed. Globalisation has led to extensive travel for vacations and work and the migration and integration of populations. In addition, people with disabilities, diseases and of different physical, mental and cultural attributes all have requirements for thermal comfort. Any provision for thermal comfort in the 21st century, therefore, should not assume that people in thermal environments all come from the same homogeneous population. The challenge for the environmental designer is to ensure that a range of different people who will occupy the 'same' environment can achieve thermal comfort.

Fanger (1970) considers the question of whether the Predicted Mean Vote (PMV) thermal comfort index is universal to all people. He investigated national geographic location, age, gender, body build, menstrual cycle and ethnic differences of people and concluded that the PMV index was valid for all populations. No differences of practical engineering significance were found in preferred comfort temperatures. McIntyre (1980) supports this noting "...Fanger has shown time and time again that the preferred temperature of a group of people does not vary and is not affected by the age, race, country of domicile, seasonal or thermal experience."

THERMAL COMFORT AND GENDER

Fanger (1970) concluded from a review of literature and experiments in Denmark and the USA up to that time, that there are only small differences, if any, between male and female subjects' comfort responses. That is when wearing the same clothing insulation. This was also shown by Yaglou and Messer (1941 – cited by McIntyre, 1980) when males and females exchanged clothing and as a consequence exchanged 'neutral' temperatures. Differences between male and female clothing will create different requirements for comfort. He also concluded that females may show more sensitivity away from comfort conditions but that there was no effect of the menstrual cycle on thermal comfort requirements (Fanger, 1970). The hormonal and other physiological and psychological changes during the menopause, as well as general belief, indicates that thermal responses will change and that there may be different requirements between menopausal and non-menopausal women. There is no conclusive evidence however, and further research is needed.

Parsons (2014) reports experimental findings that there is no difference in thermal sensation and comfort for 'neutral' and warm conditions but when environments become cold, females show more dissatisfaction than males. This was attributed to females reporting cold hands, probably because of having smaller hands and thinner fingers than males, leading to greater local heat loss. Similar results have been found in a review by Karjalainen (2012) and Yasuoka et al. (2015) and Hashiguchi et al. (2010) in Japan. There is some evidence that females are more dissatisfied than males in both warmer and cooler than comfortable environments. In addition, females are more likely to react and adapt to their environment (open windows, etc.) than males (Brager et al., 2004; Wong et al., 2009). In some unpublished experiments, I found some evidence that Chinese females were less likely to complain of discomfort (unacceptable or dissatisfied) than Chinese males or UK males and females. Reporting dissatisfaction is, however, more complex than rating thermal comfort, and there was no implication or evidence that comfort conditions differed between the four groups. These findings would need to be confirmed in a more formal study. Lan et al. (2008) found that Chinese females showed greater dissatisfaction for conditions away from comfort but that neutral temperatures were similar between males and females.

Complaints about cool air-conditioned environments among females are probably caused by maintaining air temperatures below those for comfort and differences in clothing insulation and design, where in some cases, there is a greater relative surface area of exposed skin in some female clothing. This has led to some cases of possible gender bias in the workplace where air conditioning provides cool environments and hence is less suitable for female workers.

Gerrett et al. (2014, 2015) found that females had greater thermal sensitivity to hot and cold stimuli and greater variation across the body than males. This may be an explanation for the greater adaptive response of females reported by Brager et al. (2004) although there may also be cultural reasons. I recall a US experimenter explaining that he always used female subjects when testing clothing as they were more discerning and sensitive to differences than males.

THERMAL COMFORT AND AGE

Coull (2019) produced and compared body maps and sensitivity across the body between young and elderly males and found a reduction in sensitivity with age. However, despite deterioration in thermoregulatory capacity from young to older adults for practical purposes there appears to be no difference between the conditions that we require for thermal comfort throughout our lifetime. That is if the thermal comfort conditions are specified in terms of the interaction of air temperature, radiant temperature, humidity, air velocity, clothing and activity. Adaptive opportunity will, however, vary especially as previous experience as well as physical and mental capacity will be affected by age.

New born babies just after birth are wet with great potential for heat loss due to evaporation. Babies, in general, have large head sizes relative to their bodies, large whole-body surface area to mass (volume) ratios, high skin blood flow, immature thermoregulatory systems, brown fat and hence a great capacity for heat loss. All new born babies benefit from care to maintain an acceptable thermal condition and mother's body heat in cuddling will play a role. Some babies need special care and incubator design is important (Clark and Edholm, 1985). The concept of thermal comfort is debatable in babies and any expression of mind if it exists may manifest itself in changing behaviour. Babies should be considered a special case although providing optimum conditions are of practical concern. Over-heating can be a problem due to excess clothing in an attempt to avoid over-cooling. The cultural use of swaddling being particularly dangerous.

Small children have an immature ability to maintain thermal comfort by appropriate behaviour. As they become older they often have small thin hands and feet as well as high whole-body surface area to mass ratios so are susceptible to discomfort in the cold. They often need guidance from adults in order to achieve comfort. Children frequently exhibit excessively high levels of activity and inappropriate behaviour to that which would be required for comfort.

Providing thermal comfort for school children is important as it contributes to an optimum environment for learning. There have been a number of studies, and overall there is no evidence that thermal comfort conditions for school children are different from those of adults. Adaptive opportunities and behaviour are, however, important. Wyon and Holmberg (1972) observed (via one-way mirrors) Swedish school children aged 9–11 years and found that an air temperature of 24°C was optimum. Responses to warmer environments involved changes in posture, clothing and appearance with decreased concentration (boys) and increased restlessness (girls) as pupils became hot. Humphreys (1972) used photographic records to observe 11–17 year old UK school children during the summer of 1970. Boys' uniform was long trousers, short sleeve shirt and tie; Girls' uniform was a light cotton dress. Both groups had the option to put on or take off a V-neck, long-sleeved pullover (sweater) which indicated thresholds for 'too hot' or 'too cold'. He found optimum temperatures of 24.3°C without a pullover and 21.3°C with a pullover.

Fanger (1970) cites the work of Partridge and Maclean (1935) who found no differences between adults and 7–14 year old Canadian schoolchildren. He concludes that the comfort equation (Fanger, 1970) applies for all adults but that further research is needed for children. Since that time there have been numerous studies of schoolchildren across the world and a general conclusion is that recommendations for adults are appropriate for those in schools. Adaptive behaviour and opportunity, however, has not been studied in detail and requires further research.

Fanger (1970) reviewed the literature on thermal comfort and age and also reported on climatic chamber experiments in Denmark and the USA. Comparing college-aged students with elderly people he found that there was little difference in response with elderly people preferring slightly higher air temperatures for comfort. He attributed this to elderly people having lower metabolic rates than younger people. He also noted that in practice there will be differences in lifestyle and clothing.

Collins and Hoinville (1980) found in climatic chamber studies that there was no difference between comfort conditions reported by the elderly and younger subjects. Rohles (1969) also found no differences in a survey of 64 adults over the age of 70 years when compared with comfort requirements of college-aged students. Lankilde (1977) also found similar results for people over 80 years of age. Cena and Spotila (1986) in an extensive survey on thermal comfort concluded that for elderly people in the USA, there were no significant effects of gender, age, health, smoking, alcohol usage, income or house value on thermal comfort conditions. They conclude that the PMV index has validity for elderly people.

Wong et al. (2009) also found that the PMV index was a reasonable predictor of discomfort conditions in a survey of 384 people aged 60–97 years old across 19 nursing homes. This is reported in ISO TR 22411 (2019) (from Japanese standard JIS TR 0002:2006) where the percentage of older people reporting warm or cold to the same thermal stimulus is less than that for younger people. It is also reported that older people were less likely to find an environment 'unacceptable'.

There may be differences between comfort requirements for people of different ages but much of these are related to metabolic heat production and clothing which are variables included in (for example) the PMV comfort index. Larger differences may be found in moderate to more extreme environments as thermoregulatory capacity and perception of heat and cold changes with age. Of particular importance will be the adaptive opportunities available to and acted upon by people of different ages. This will be an area for fruitful research in the future.

Van Hoof et al. (2019) have reported on a qualitative approach to studies of thermal comfort in the elderly. Using focus groups they investigated how older people achieve thermal comfort in homes. They identified four factors: personal factors (age, gender, financial position, etc.), knowing (past experience and knowledge of thermal environments and what affects comfort, etc.); feeling (thermal sensation, comfort, etc.) and doing (adaptations, technological solutions, etc.) as how older South Australians deal with the thermal challenges of living environments. This approach has the potential to provide the basis for linking accepted thermal comfort models with appropriate environmental design.

A final word on thermal comfort for the elderly is that they have entered the phase of their lifetime that, on average, begins with relative health and accelerates towards morbidity and eventual mortality. Requirements become more and more specific to the individual as mechanisms for deterioration of health between (and within) individuals as well as possible cures and mitigations (including drugs) diverge. Almost by definition, those people who have been investigated for thermal comfort requirements, particularly in laboratory studies, are self-selected and healthy in the sense that they are capable and acceptable for the studies. They therefore represent a biased sample and subset of people as they age from around 60–65 years old to death at on average around 80–85 years old depending upon gender, country of origin, social status, affluence, social interaction, lifestyle and more.

NATIONAL GEOGRAPHIC LOCATION

McIntyre (1980) cites the work of Fanger (1970), noting that wherever people come from, their mean preferred temperature will be 25.6°C ± 1.6°C for the same standard conditions of rest and light clothing. Fanger (1970) reports a climatic chamber experiment to determine the temperatures required for thermal neutrality by Danish college age subjects and to compare them with those determined by Nevins et al. (1966) in climatic chamber studies at Kansas State University in the USA. He found no statistical difference between the groups with a mean difference of less than 0.2°C. He concluded that as his proposed comfort equation agreed very well with the results, it will be valid for people in the USA, Denmark and across temperate climate zones. He also concluded that any differences found in field studies can be attributed to variations in clothing and not due to adaptation but due to differences in outdoor climate or way of life.

Although at the time (1970) thermal comfort in other regions such as the tropics, required further research, much has now been conducted. Parsons (2002) reported a laboratory study of thermal sensation responses of subjects before and after an acclimatisation (acclimation) program. For neutral and slightly warm to warm conditions, no difference in sensation was found between subjects when they were unacclimatised and when they were acclimatised. Being acclimatised to heat therefore allows greater resistance to heat stress but does not alter thermal comfort requirements. Fanger (1970) analyses the data from studies conducted in Calcutta, Singapore, Nigeria, North Australia and New Guinea and shows a remarkable agreement with the neutral temperatures predicted by the PMV index and comfort equation.

There have been many studies into whether the PMV/PPD (Predicted Percentage of Dissatisfied) indices are valid throughout the world. These have been over all continents and particularly with Japanese and more recently Chinese people. Tanabe et al. (1987) conducted thermal comfort climatic chamber studies on Japanese subjects that were similar to those conducted by Nevins et al. (1966) in the USA. He found close agreement between the two studies and concluded that the PMV/PPD indices were valid for use in Japan. Li and Lim (2013), Parsons (2014) and Yao et al. (2009) have found similar results for Chinese students in studies in the UK and China. There are now numerous other examples and a consistent result that thermal comfort conditions do not vary with national geographic location. There is, however, evidence that preferred temperatures do depend upon adaptive opportunities. That is not that the thermal comfort requirements in terms of the interaction of the effects of air temperature, radiant temperature, humidity, air velocity, clothing and activity are different for different populations, but that the method of achieving optimum conditions varies due to the use of a variety of adaptive opportunities to achieve adaptive thermal comfort.

For the same thermal conditions, for people across the world, there will be differences in reported discomfort, dissatisfaction, acceptability, complaints and tolerance. This will greatly depend upon the psychological and social climate. While comfort conditions will not vary, the threshold of complaint, for example, will depend greatly on likelihood of change, possibility of being chastised, importance and motivation to carry out tasks, reward and other factors. An interesting finding is in the definition of thermal comfort as 'neutral'. For people who live in hot climates, mostly attempting to keep cool, the term 'cool' may have positive connotations. In contrast, people in cool climates may favour the term 'warm' (Tochihara et al., 2012).

PEOPLE WITH DISABILITIES

People with mental disabilities have not been studied in detail. It is reasonable to assume that maintaining thermal comfort conditions is important and that the ability to adapt to thermal conditions is reduced. As a first approximation, it seems likely that the best starting point for environmental design is to establish thermal conditions comfortable for those without mental disabilities.

In addition, assistance may be needed from people without mental disabilities to ensure behaviours are appropriate to maintain comfort. This will be specific to individual requirements as each person will be unique and there will be a great diversity of characteristics. This was also the first conclusion of the research of Webb and Parsons (1997) in their extensive studies of thermal comfort for people with physical disabilities. A case study approach was recommended as it was found that each individual person with a physical disability had a number of unique characteristics and requirements in addition to their simple classification in terms of their disability (e.g. it was found that a person with cerebral palsy may also exhibit epilepsy, side effects of drugs and more). It was also concluded that adaptive opportunities were important and greatly differed from those of people without physical disabilities. It was essential that they were given prime consideration. A person who cannot adjust clothing or move away from a draught due to restricted physical capability requires specific attention in environmental design to allow the achievement of thermal comfort.

Giorgi et al. (1996) conducted a review of research into the thermal comfort requirements for people with physical disabilities and found many confounding factors and no general conclusions. Yoshida et al. (2000) report on a survey of five special schools for pupils with severe disabilities in Japan. Among 87 children, 67 suffered from major disorders of thermoregulation. It was concluded that teachers and family members were required to maintain optimum thermal conditions for those children and that much more work on environmental design for such children was required.

Parsons and Webb (1999) and Parsons (2014) report on an extensive series of climatic chamber studies involving 531 subject trials in 'cool', 'neutral' and 'warm' environments, and field studies involving 391 people with physical disabilities at home and other residential buildings and 38 carers in care homes and holiday centres. The field studies involved questionnaires and elicited a wealth of data on existing thermal discomfort and restrictions to adaptive opportunity to achieve comfort. There was a general feeling that people with physical disabilities require higher temperatures than those without physical disabilities for comfort, although this may be due to a comparison between active carers and sedentary people with physical disabilities.

The laboratory studies involved people with cerebral palsy, spinal injury, spinal degeneration, spina bifida, hemiplegia, polio, osteoarthritis, rheumatoid arthritis, head injury and multiple sclerosis (MS). It was found that on average the PMV index was a good first approximation to the average thermal sensation of people with physical disabilities although the variation in response was generally greater than that of people without physical disabilities. A questionnaire study of care and residential homes and particularly day centres (27 of 38) answered by 38 cares with a mean of 12.64 (±9.48 SD) years experience indicated that most complaints are for people who want to be warmer. The carers usually responded by informing the caretaker (janitor) or adjusting the heating themselves. People with disabilities generally wore the same level of clothing or sometimes more than those without disabilities and their thermal conditions were greatly influenced by others. The results seem to confirm that thermal conditions are often determined by people who are active when people with disabilities are often inactive. To some extent, this is also confounded by age. For more details of these studies, see Parsons and Webb (1999) and Parsons (2014). A summary of the laboratory results is shown in Figure 8.1.

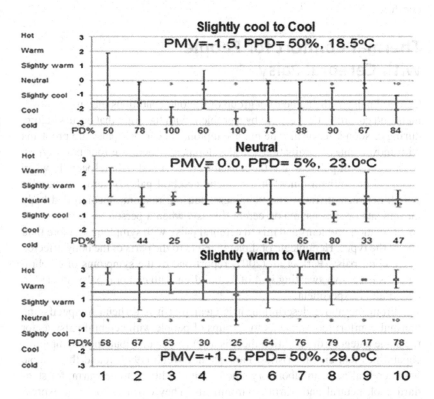

FIGURE 8.1 Mean and standard deviation of thermal sensation responses and percentage of dissatisfied (PD) for ten groups of people with physical disabilities exposed to PMV = −1.5, slightly cool to cool; PMV = 0, neutral; and PMV = +1.5, slightly warm to warm conditions. Groups: 1. Cerebral palsy (N = 12); 2. Spinal injury (N = 9); 3. Spinal degeneration (N = 8); 4. Spina bifida (N = 10); 5. Hemiplegia incl stroke (N = 4); 6. Polio (N = 11); 7. Osteoarthritis (N = 17); 8. Rheumatoid arthritis (N = 10); 9. Head injury (N = 6) and 10. Multiple sclerosis (N = 32).

A description of each of the categories of disability investigated in the study by Parsons and Webb (1999) is provided below with a summary of results. An attempt is made to identify possible effects on adaptive opportunity. These are interpretative based upon the descriptions of the disability. The descriptions are important, however, further investigation is required. A good starting point for considering restrictions to adaptive opportunities is to consider limits on the ability to move (and move around), adjust clothing and use technical methods for creating thermal comfort conditions. Technical aids to compensate for the effects of a disability, such as wheelchairs and callipers, will influence (aid or restrict) the adaptive opportunities to achieve thermal comfort.

Thermal Comfort for People with Cerebral Palsy

Cerebral palsy is the name for a group of lifelong conditions that affect movement and co-ordination, caused by a problem with the brain that occurs before, during or soon after birth. The main symptoms of cerebral palsy are problems with movement, co-ordination and development. There are four main types: spastic cerebral palsy, where the muscles are stiff and tight (especially when trying to move them quickly); dyskinetic cerebral palsy, where the muscles switch between stiffness and floppiness, causing random, uncontrolled body movements or spasms; ataxic cerebral palsy, when a person has balance and co-ordination problems and mixed cerebral palsy with symptoms of more than one of the types. Hemiplegia or diplegia refers to the parts of the body affected by cerebral palsy. Park et al. (2002) report one of the symptoms of cerebral palsy as excessive sweating in 4 of 12 children with cerebral palsy studied in a laboratory experiment.

From the above description, it is clear that it is difficult to specify the thermal comfort requirements for groups of people with cerebral palsy as for the same category of condition there is a very wide distribution of symptoms, capabilities and characteristics. Parsons and Webb (1999) found that people with cerebral palsy in laboratory studies felt conditions to be warm, for standard cool, neutral and warm environments. They did not, however, express greater dissatisfaction than expected for people without physical disabilities. From questionnaire studies in day centers and residential homes, carers indicated that in their experience, people with cerebral palsy wanted to be warmer than people without disabilities and wore about the same level of clothing.

Adaptive opportunities for people with cerebral palsy when compared with those without disabilities will be restricted because of the disability.

Clothing adjustment, being able to move around and opportunities provided by technology if not appropriately designed (e.g. operating controls) may all be reduced depending upon the type and severity of cerebral palsy.

Thermal Comfort for People with Spinal Injury

Spinal cord injury (SCI) is an insult to the spinal cord resulting in a change, either temporary or permanent, in the cord's normal motor, sensory, or autonomic function. Patients with spinal cord injury usually have a multisystem impairment, which is often permanent and threaten health, function and social participation. When caring for a person with a spinal cord injury, it is important to look at them holistically. There are two types: traumatic, due to a fall or injury, for example, and non-traumatic, due to tumours, inflammation or degeneration. The part of the spinal cord that was damaged corresponds to the spinal nerves at that level and below. Injuries can be cervical 1–8 (C1–C8), thoracic 1–12 (T1–T12), lumbar 1–5 (L1–L5) or sacral (S1–S5). A person's level of injury is defined as the lowest level of full sensation and function. The majority of people who sustain spinal cord injuries are not able to sweat below the level of injury. For people who sit all the time, blood pools into the feet in legs and legs can become cold even in hot weather.

Parsons and Webb (1999) found that people with spinal cord injury (including people with tetraplegia and quadriplegia) provided similar thermal sensation votes to those without disabilities as predicted by the PMV values (ISO 7730, 2005). As conditions become warm or cool variation in response greatly increased. Dissatisfaction levels, however, were much greater than predicted for those without physical disabilities. From questionnaire studies in day centers and residential homes, carers indicated that in their experience, none of the people with tetraplegia and quadriplegia wanted to be cooler than people without disabilities and wore about the same level of clothing.

Adaptive opportunities, when compared with those without disabilities, will be restricted because of the disability. Clothing adjustment, being able to move around and opportunities provided by technology if not appropriately designed (e.g. open windows) may all be reduced depending upon the type and severity of the spinal injury.

Advice on adaptive opportunities for people with spinal cord injury is provided by Alicia Reegan, a person with spinal cord injury (Reegan, 2017). Alicia gives a few preventative tips on how to cool your body down: carrying around a spray bottle to create artificial sweat; placing an ice cold towel around your neck; sitting in the shade; sitting in your car with air-conditioning blasting; attempt to move your legs around periodically throughout the day; dress in layers; heated blankets; taking a hot shower; sitting in your car with the heater blasting. The advice emphasizes the power of qualitative methods in appropriate environmental design. It would seem essential to include the user with the disability in the design or if not available then to gain inspiration from focus groups and other information gathering methods.

Thermal Comfort for People with Spinal Degeneration

Degenerative spine conditions involve the gradual loss of normal structure and function of the spine over time. They are usually caused by aging but may also be the result of tumours, infections or arthritis. Pressure on the spinal cord and nerve roots caused by degeneration can be caused by slipped or herniated discs.

Parsons and Webb (1999) found that people with spinal degeneration are cooler than expected in slightly cool to cool environments (PMV = −1.5) and warmer than expected in slightly warm to warm conditions (PMV = 1.5). Conditions for neutrality (PMV = 0) are as expected. There is greater percentage dissatisfaction reported for all conditions than would be expected for people without disabilities with all eight subjects dissatisfied in the cool environment. Adaptive opportunities will depend upon the extent of the condition. Ability to move around will be restricted as well as to adjust clothing. Technical solutions will require special consideration.

Thermal Comfort for People with Spina Bifida

Spina bifida is when a baby's spine and spinal cord don't develop properly in the womb, causing a gap in the spine. Spina bifida is a type of neural tube defect, which is the structure that eventually develops into the baby's brain and spinal cord. In severe cases the baby's spinal canal remains open along several vertebrae forming a sac in the back. In less severe cases the gaps are small and don't cause problems.

Parsons and Webb (1999) found that people with spina bifida were on average (n = 10) slightly warm in neutral environments (PMV = 0), neutral in slightly cool to cool (PMV = −1.5) environments and warm in slightly warm to warm (PMV = 1.5) environments with a large variation in response. There was slightly more dissatisfaction than expected in neutral and cool environments but less dissatisfaction in warm environments.

Thermal Comfort for People with Hemiplegia

Hemiplegia is a condition that affects one side of the body (right or left). It is caused by injury to parts of the brain that control movements of the limbs, body, face, etc. This injury may happen before, during or soon after birth. If this happens later in life as a result of injury or illness, it is called acquired hemiplegia.

Parsons and Webb (1999) found that on average people with hemiplegia were cold in cool conditions and cool in neutral conditions and slightly warm in slightly warm conditions but with a large variation. They were dissatisfied with neutral and especially cool environments but less than expected in warm environments. It is clear from the description of the condition that adaptive opportunity will be restricted and will require special consideration.

Thermal Comfort for People with Polio

Polio, *short for poliomyelitis*, is a serious viral infection that used to be common in the UK and worldwide. It's rare nowadays because it can be prevented with vaccination. In some cases, there is muscle weakness resulting in an inability to move. This can occur over a few hours to a few days. The weakness most often involves the legs but may less commonly involve the muscles of the head, neck and diaphragm. Complications after recovery include skeletal deformations and hence difficulty in movement often requiring supports to the limbs. Although polio often passes quickly without causing any other problems, it can sometimes lead to persistent or lifelong difficulties. A few people with the infection will have some degree of permanent paralysis, and others may be left with problems that require longterm treatment and support. These can include: muscle weakness; shrinking of the muscles (atrophy); tight joints (contractures); and deformities, such as twisted feet or legs.

Parsons and Webb (1999) found that people with the effects of polio showed greater dissatisfaction than expected in cool, neutral and warm environments. They were warm to hot in slightly warm environments, slightly cool in slightly cool environments, and neutral in neutral environments. There was much variation about the mean however. People with the effects of polio were reported by carers to wear the same clothing as people without disabilities. It is clear that if the person who has had polio has mobility problems, then adaptive opportunity will be restricted.

Thermal Comfort for People with Osteoarthritis

Osteoarthritis is a condition that causes joints to become painful and stiff. Some people also experience swelling, tenderness and a grating or crackling sound when moving the affected joints.

Parsons and Webb (1999) found that people with osteoarthritis were on average cooler than expected in the cool environment, warmer than expected in the warm environment and around neutral in the neutral environment. There was large inter-subject variation and much dissatisfaction in all conditions. Carers confidently note that people with arthritis require warmer temperatures for comfort than people without disabilities. Relative to a healthy person without disability, adaptive opportunity will be reduced due to restricted mobility and pain. Carers note that over half of the people in their care with arthritis require more clothing than people without disabilities and none require less clothing.

Thermal Comfort for People with Rheumatoid Arthritis

Rheumatoid arthritis is a long-term condition that causes pain, swelling and stiffness in the joints. The symptoms usually affect the hands, feet and wrists. It is often considered to be exacerbated by cold, high humidity and causes a high internal body temperature (fever) and sweating.

Parsons and Webb (1999) found that as for people with osteoarthritis, people with rheumatoid arthritis were cooler in cool environments and warmer in warm environments with much variation. However, they were slightly cool in neutral environments with much dissatisfaction in all conditions. Carers suggested that they require to be at warmer temperatures than people without arthritis.

Thermal Comfort for People with Head Injury

Symptoms of a severe head injury can include: fits or seizures; difficulty speaking or staying awake; problems with the senses such as hearing loss or double vision; repeated episodes of vomiting; blood or clear fluid coming from the ears or nose; memory loss; sudden swelling or bruising around both eyes or behind the ear; difficulty with walking or co-ordination. Some are short-term symptoms and others long term.

Parsons and Webb (1999) found that people who had had a head injury were on average warmer in warm environments but close to neutral in cool and neutral environments. There was great individual variation for cool and neutral conditions but not for warm environments where there was much agreement between subjects that they were warm. There was lower than expected dissatisfaction in the warm environment but larger dissatisfaction in neutral environment (33%) and much dissatisfaction in the slightly cool to cool environment (79%) see Figure 8.1. Clearly the restriction to movement will reduce the opportunity to adjust clothing and move around to achieve thermal comfort.

Thermal Comfort for People with Multiple Sclerosis

Multiple sclerosis (MS) is a condition that can affect the brain and spinal cord, causing a wide range of potential symptoms, including problems with vision, arm or leg movement, sensation or balance. MS is a lifelong condition that can sometimes cause serious disability, although it can occasionally be mild. It is most commonly diagnosed in people in their 20s and 30s (and in northern cooler climates) although it can develop at any age. It is about 2–3 times more common in women than men. MS is one of the most common causes of disability in younger adults. The symptoms of MS vary widely from person to person and can affect any part of the body. The main symptoms include: fatigue; difficulty walking; vision problems, such as blurred vision; problems controlling the bladder; numbness or tingling in different parts of the body; muscle stiffness and spasms; problems with balance and co-ordination; problems with thinking, learning and planning.

Parsons and Webb (1999) and Webb et al. (1999) found that people with MS were on average cooler in the cooler environment and warmer in the warm environment than expected for people without disabilities according to ISO 7730 (2005) although there was a large inter-subject variability. People with MS were neutral in neutral environments with low variation but significant dissatisfaction (47%). Roughly equal numbers of carers thought people with MS would like to have cooler conditions and warmer conditions than those without physical disabilities but few thought that they would prefer the same conditions.

OTHER POPULATIONS WITH SPECIAL REQUIREMENTS

In recognition that ISO Standards for thermal comfort had been designed mainly with data from investigations into non-disabled, healthy adults and mainly college age students, ISO TR 14415 (2005) was produced to provide information on thermal comfort conditions for people with disabilities. This was later superseded, updated and extended into ISO 28803 (2012) to cover physical environments in general. From the perspective of thermal environments sensory impairment and paralysis, differences in shape of the body, impairment of sweat secretion; impairment of vasomotor control; differences in metabolic rate and effects of psychological stress were identified as the characteristics of people with special requirements (outside of the scope of other thermal comfort standards) that would differ from people without special requirements and be relevant to human response to thermal environments and hence thermal comfort. The standard suggests that the existing thermal comfort standard (ISO 7730, 2005) using the PMV and PPD indices will provide acceptable predictions for neutral conditions but will show wide variation between people. Also that the equations for sweat rate and skin temperature for comfort may vary with special populations and these have not been validated.

There are many ways in which people can be categorised, and further research is clearly required into individual factors. Fanger (1970) concludes that there is no significant difference between the comfort requirements of obese and thin adults. In the absence of studies with human subjects, logical considerations suggest that obese people will have much greater metabolic rate in activities involving walking uphill, for example, this would apply to pregnant women where hormonal changes may influence comfort conditions and sick people will have specific requirements relating to their illness and medication. All of these factors require further investigation both in terms of whether existing thermal comfort indices are valid and what influence they have on adaptive opportunities.

Thermal Comfort and Human Performance

9

THE HSDC METHOD

After much investigation into the effects of the thermal environment on the ability of people to carry out tasks and jobs, it is reasonable to conclude that optimum conditions for human performance are those that also provide thermal comfort. This is consistent with the definition of thermal comfort as a condition providing no strain on the body as no change is required and hence does not utilise resources in terms of preserving health, attention to a task and any loss in ability to perform tasks. There have been many studies of the effects of the thermal environment on human performance and a practical method of prediction is to consider that there are three main effects. These are due to Health and Safety (HS); Distraction (D) and Capacity (C) (Parsons, 2018, 2019; ISO NWI 23454 Part 1, 2019).

The Health and Safety, Distraction and Capacity (HSDC) model of human performance is a method for predicting the effects of the environment on human performance and productivity. It involves calculating the multiple of; the percentage of worktime (e.g. over a shift) available for safe work (HS); the percentage of time on work not distracted by the environment (D); and the capacity to perform work or a task when compared with the capacity under optimum conditions (C). That is, the reduction in performance at a task or work when compared with optimum performance under comfort conditions is HS × D × C.

Suppose that people are working in an office at 27°C and they are slightly distracted due to dissatisfaction as they are too warm (e.g. Predicted Mean Vote, PMV = +2, 50% dissatisfied) such that they attend to the environment rather than the task for 10% of the time (e.g. assuming that one in five of those dissatisfied will be distracted). On average, people are therefore 10% distracted and hence working on the task only 90% of the time (D = 0.90). If there is evidence that the capacity to carry out the task when warm is only 95% of that when comfortable, then capacity is 95% (C = 0.95). As this is an office environment and comfort and not health and safety are of concern, there will be no loss in time off work due to health and safety or other regulations. Therefore, time available for work is 100% of the time available for this moderate environment (HS = 1). The level of performance in this environment is predicted to be HS × D × C, that is, 1 × 0.90 × 0.95 = 85.5% of the performance in optimum comfort conditions. For a highly motivated person or group, there may be no distraction due to the environment; and for some tasks, there will be no loss of capacity for such a small deviation from comfort conditions. The prediction of HS × D × C will then be 1 × 1 × 1 = 100% (i.e. the same performance as for comfortable environments). Although the HSDC method assumes optimum conditions for comfort in theory, it could be greater than 100%. It is also useful to consider adaptive thermal comfort in this context.

Although the comfort conditions will remain as optimum conditions, there may be some strain and distraction due to adaptive behaviour and so adaptation itself may have a cost to performance. Despite the obvious advantages in versatility of environmental design using adaptive approaches, providing comfort conditions without the need for adaptive behaviour may therefore have some advantages in the provision of optimum conditions for human performance.

THERMAL COMFORT AND HEALTH AND SAFETY (HS)

If people become so uncomfortable due to thermal conditions, that they refuse to work or if regulations lead to a cessation of work as thermal limits are exceeded, then performance and productivity will fall to zero in times when people are not working (e.g. for all or part of a day). This is unlikely to occur in moderate environments and where thermal comfort and not health is a main consideration. However, towards the uncomfortable extremes of thermal environments this may occur. In the UK, the factories, shops and railway premises (factories) act (1961) stipulated that air temperatures at the workplace must be at 16°C or more for sedentary office work, or 13°C or more for active work, after 1 hour of starting work. There are generally no upper limits for work environments, however workers have been known to 'walk out' due to unacceptable hot conditions, schools are sometimes closed, etc. An upper limit of 33°C for air temperature was recommended at the start of this book and could be used as a realistic value for calculations of loss in performance.

DISTRACTION (D)

Parsons (2014) defines thermal distraction as "...a tendency for a person to attend to a thermal state (hot, uncomfortable, cold) instead of performing a task". If distracted, therefore performance at the task will be zero for the time of the distraction. There are a number of possible causes of distraction such as attention to other stimuli and tasks than the ones being performed; however for the purposes of the effects of the environment on performance, it is the distraction caused by the effects of the environment that are of interest; and for thermal comfort, it will be related to the degree of discomfort, which may lead to dissatisfaction, that requires attention.

The relationship between dissatisfaction due to thermal discomfort and distraction has not been established and will depend upon a number of contextual and other factors. We can consider that dissatisfaction is a driver for distraction and that whether someone is distracted or not will be related to how motivated the person is to perform the task or work of interest. Thermal comfort indices such as the PMV (Fanger, 1970) predict the mean thermal sensation of a group of people from thermal conditions made up of the air temperature, radiant temperature, humidity, air velocity, clothing and activity. From the PMV the Predicted Percentage of Dissatisfied (PPD) is derived. It would seem reasonable that dissatisfaction leads to distraction and that this will depend upon the disposition of the person to respond to their thermal state and not to the task in hand. This relationship is not fully understood but if we assume that for people of normal disposition, motivated to perform the task, one in five of the people dissatisfied will be inclined to do something about it and hence be distracted, we can predict the PPD from the thermal conditions and hence the percentage of time a person will be distracted. If the PMV is −2, cold, the PPD is 50%, so 10% of people will be distracted. If we assume 10% of the time people are distracted, then we can say that in this cold environment performance will be only 90% of what it would be if the conditions were comfortable where any distraction would be zero according to the rationale (or close to zero by calculation as for PMV = 0, PPD = 5% so D = 1%). For the purposes of this discussion the assumption is that this will be the best prediction of loss in performance for both a group and an individual. It is likely that rapid and unexpected or undesirable changes in the perception of thermal conditions, such as those in transient environments, will cause greater distraction, however there have been no studies on this phenomenon.

For highly motivated people or people who are not inclined to respond to, and hence will accept, dissatisfaction, there will be no or little distraction and hence no loss in performance due to distraction.

CAPACITY (C)

There is no doubt that thermal environments can affect the capacity of people to perform tasks, however, despite many laboratory and field studies, the establishment of any predictive model has been elusive. A first step in the process is to define what we mean by human performance and this is usually done with a task taxonomy. The first obvious stage is to categorise tasks as manual or cognitive. Ramsey and Kwon also include perceptual-motor tasks. Parsons (2014) summarises results including the extensive reviews of Ramsey and Kwon (1988) and the experiments of Meese et al. (1984). Wyon et al. (2001) summarises the effects of thermal environments on performance (capacity to carry out the task) as they deviate from thermal neutrality; and suggests that personal control of the environment (using workplace individual controls) allows ranges of conditions to be set that will allow optimisation of performance for individuals.

A common experience for those investigating the effects of the thermal environment on the capacity and hence performance at tasks is inconsistency in results. That is with the probable exception of loss in manual dexterity as hands become cold. Many tasks have been studied and it is typically found that, for identical thermal and contextual conditions, in some studies there is a detriment in performance, in some an improvement and in some no change. This was found in the extensive review of Ramsey and Kwon (1988) and it can reasonably be concluded that there is little loss in capacity to perform tasks in moderate conditions. The HSDC model proposed above for all environments would therefore focus on distraction as the major possible driver to loss in performance. Ramsey and Kwon (1988) proposed a similar model describing task performance, arousal, physiological distraction and perceived control as the main factors. It is logical to speculate that thermoregulatory response will influence task performance. Vasodilation may increase speed of response, and vasoconstriction may lower it as well as manual dexterity. Sweating hands will affect grip. It is reasonable, however, to conclude that if moderate environments affect the capacity to carry out tasks it has yet to be formulated in a consistent way.

THE ISO STANDARDS INITIATIVE

Parsons (2018) describes an initiative to develop a series of international standards concerned with specifying the effects of the physical environment on human performance. The first of these is proposed as ISO NWI 23454 – 1 (2019): Human performance in physical environments: Part 1 – A performance framework. The strategy is to produce two generic standards that can be used to describe the effects of specific environmental components (thermal, noise, light, etc.). The first generic standard will propose a model of human performance. This is shown in Figure 9.1. It will also describe a method for predicting performance based upon the HSDC method. The second will describe standardised tasks that can be used to measure performance at the components of the model. (ISO DP 23454 – 2 (2019): Human performance in physical environments: Part 2 – Measures and methods for assessing the effects of the physical environment on human performance).

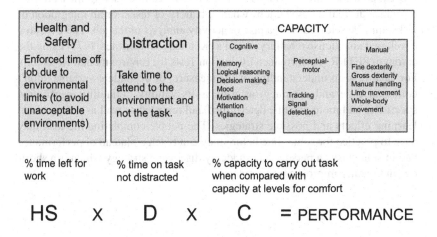

FIGURE 9.1 HSDC model for predicting the effects of the environment on human performance.

CAPACITY AND THE ISO PERFORMANCE MODEL

Due to inconsistencies in experimental results and a lack of agreed philosophy about what is meant by human performance, a task taxonomy is never perfect. An ISO performance model, however, will provide consistency and direction for future studies of the effects of the environment on human performance. It will also provide a way forward and promote understanding in the subject. The proposal in the Technical Report is to divide human performance into three areas of performance – cognitive, perceptual-motor and manual (sensory was also considered but this is an integral part of the other areas), with the following sub-divisions: *Cognitive performance* (attention; vigilance; signal detection; learning; logical reasoning; memory; decision making); *Perceptual-motor performance* (tracking); *Manual performance* (fine dexterity; gross motor performance; lifting and handling; endurance).

Examples of standard tasks (there are many and some commercially produced) would include manipulation of pins in holes (testing manual dexterity) or for cognitive assessment, answering a set of logical reasoning questions (e.g. if A > B and B > C, then A < C, True or False?). There are many more that can be used to assess performance (capacity as affected by environmental conditions). Different tasks can be identified to measure capacity at each of the components of the performance model. Where a number of tasks are used it is called a test battery and a selection of tasks can be used to measure performance at 'real' complex jobs where a battery of tasks, or an integration of tasks, may be selected from a task or activity analysis of the job. The US Navy developed an extensive battery of tasks (Kennedy and Bittner, 1977) using the acronym PETER (Performance Evaluation Tests for Environmental Research). For each task an understanding of the measure, or dependent variable, (don't confound time and accuracy) as well as its validity, reliability, sensitivity, specificity and more must be determined and reported as well as details about learning effects and possible strategies. The people completing the task may have to practice for some time before they can reach a plateau of performance and in some tasks a person may suddenly discover a strategy which is a short-cut and greatly improves performance.

International Standards and a Computer Model of Thermal Comfort

10

INTERNATIONAL STANDARDS AND THERMAL COMFORT

The two most influential standards used internationally to specify thermal comfort conditions, mainly, but not only, for buildings across the world are ISO 7730 (2005) "Ergonomics of the thermal environment – Analytical determination and interpretation of thermal comfort using calculation of the PMV and PPD indices and local thermal comfort criteria" and ASHRAE 55 which is reviewed (but not necessarily revised) annually. For 2017, it is ASHRAE 55-2017 "Thermal Environmental Conditions for Human Occupancy (ANSI/ASHRAE Approved)". As an interim action, addenda to the standard are published annually. Addendum for 2019, for example, provides an updated computer program in the computer language Java-script, for calculating the Predicted Mean Vote (PMV)/Predicted Percentage of Dissatisfied (PPD) indices. ASHRAE 55 presents comfort limits in terms of air temperature and humidity on a psychrometric chart. It also presents an adaptive approach to achieving thermal comfort (for a fuller discussion, see Chapter 6, and Nicol et al., 2012). Comfort zones are provided on the psychrometric chart of operative temperature (t_o °C – a weighted average of mean radiant temperature and air temperature) and dew point (dp °C – the temperature at saturated water vapour pressure when water will condense out of the air), which is related to humidity. For light activity the zones of thermal comfort are given as areas bounded by lines on the psychrometric chart (see Figure 6.2). ASHRAE provides the following answer to the frequently asked question, What are the recommended indoor temperature and humidity levels for homes? "…temperatures could range from between 67°F (19.5°C) and 82°F (27.8°C) but depends upon relative humidity, season, clothing worn, activity levels and other factors. …HVAC (Heating, Ventilation and Air Conditioning) systems must be able to maintain a humidity ratio of at or below 0.012 … an upper relative humidity level of more than 80% … The standard does not specify a lower humidity limit."

ISO 7730 (2005) is the international standard that specifies conditions for thermal comfort and the prediction of the degree of thermal discomfort and is used by other standard bodies as the basis for standards in specific areas of application such as outdoors or in buildings. One of its main functions is to standardise the definition and calculation of the PMV and PPD indices as well as methods for predicting local thermal discomfort. It forms part of a series of international thermal comfort standards concerned with subjective assessment (ISO 10551, 2019); physical environments for people with special requirements (ISO 28803, 2012); thermal sensation caused by skin contact with surfaces of moderate temperature (ISO 13732 Part 2, 2001); the environmental survey (ISO 28802, 2012); thermal comfort in vehicles (ISO 14505 Parts 1–4, 2006) and for the practical design of environments for thermal comfort (ISO 16594, 2019 in development). There are also supporting standards specifying instruments for the measurement of the thermal environment (ISO 7726, 1998); the estimation of the thermal properties of clothing (ISO 9920, 2007) and the estimation of metabolic heat production from type of activity (ISO 8996, 2004).

ISO 7730 (2005) describes and defines how to calculate the PMV of a large group of people exposed to a thermal environment defined by the air temperature, radiant temperature, humidity, air velocity, metabolic rate and clothing. The PMV is a value on the thermal sensation scale: 3, hot; 2, warm; 1, slightly warm; 0, neutral; −1, slightly cool; −2, cool; −3, cold. A sensation of neutral (PMV = 0) is taken as conditions for thermal comfort. The PPD predicts how many people would be dissatisfied with the conditions and is calculated from the PMV value. The calculations are identical to those first described by Fanger (1970) and presented in detail in Chapters 2 and 3. There is no doubt that the PMV/PPD indices are used to best effect through a computer program. The formulations can then be used as models of thermal comfort allowing assessment and evaluations of thermal environments and specification and testing of environmental designs in 'what if' scenario testing, recommendations for change and so on. One of the great advantages of the PMV/PPD indices is that their equations have not changed since their inception and that standardisation had consolidated that position. That means whenever and wherever in the world they have been derived (correctly) for the same conditions, there will be the same outcome. This is a non-trivial point as often thermal comfort methods evolve, and it is important to know which version of a thermal comfort model has been used. Although when first proposed by Fanger (1970) and for 10 or more years after that the PMV/PPD values were derived by practitioners using tables and charts. It is now natural to use a computer program, which allows its use to full potential. For this reason a computer program is described below in detail that allows the reader to construct their own computer model of thermal comfort.

A COMPUTER MODEL OF THERMAL COMFORT

There are a numerous places 'on the internet' where versions of software to calculate the PMV/PPD thermal comfort indices can be found. Any internet search will provide a list. The reader should check that they provide valid results before using them, using sample inputs and outputs provided in ISO 7730 (2005) and in Chapter 3. It is possible for practitioners to use such software as a 'black box' without understanding of the contents of the computer program. My own view is that this is not recommended and as the procedure requires little effort I would suggest constructing your own computer program. A fuller understanding of inputs, calculations and outputs as well as sensitivity to different variables can then be derived. The calculation of the PMV/PPD indices are presented below. Other thermal comfort indices are also used, for example, the New Effective Temperature (ET*) and the Standard Effective Temperature (SET) indices. These are based upon the two node model of human thermoregulation for which a description of how to construct a computer program is provided in Parsons (2019).

PROGRAM DEVELOPMENT

There are many common features about computer programming languages, particularly when mathematical equations are used so any presentation will be easily transformed between languages. ISO 7730 (2005) provides a program listing (source code) in the language BASIC. Unfortunately in publication, important editorial errors have occurred (an obvious missing line and a couple of other typos) so I have pointed them out below and corrected for them in the version described in this chapter. A 2019 addendum to ASHRAE 55-2017 provides a version of the computer program source code in Java-script and I have used the (corrected) version in ISO 7730 (2005) in spreadsheet form (Excel ver 10.1, 2010). To start this process in the widely available Microsoft Excel, the reader should be familiar with that software and format it, so that 'Developer' is available on the main menu. This will allow programming in a visual basic macro with easy movement between developing and editing the program and running the program on the spreadsheet. If another software 'platform' is used the description below will also be valid.

CORRECTIONS TO THE SOURCE CODE IN ISO 7730 (2005)

The source code computer listing for calculation of the PMV and PPD indices is provided in ISO 7730 (2005) Annex D (normative therefore part of the standard) in the computer language BASIC. Each line of code is numbered from 10 to 580 in steps of 10. A main error is the omission of line 490 that seems to have been lost while editing. A correction is to add the line '490' (between 480 and 500) which calculates the heat transfer by convection and is given the symbol HL6 as follows.

```
490 HL6 = FCL*HC*(TCL - TA)
```

In line '140' a function is defined for saturated vapour pressure (kPa) that has a missing bracket before T + 235). It should be

```
140 DEF FNPS(T) = EXP(16.6536 - 4030.183/(T + 235))
```

Line '200' has a spacing problem and should be

```
200 If ICL ≤ 0.078 Then FCL = 1 + 1.29*ICL
```

Line '300' replace * with ^ as in

```
300 P5 = 308.7 - 0.028*MW + P2*(TRA/100)^4
```

Line '310' TLCA should be TCLA

```
310 XN = TCLA/100
```

Further corrections are in line '460', line '470' and in line '510'; lower case 'm' representing metabolic rate in W m^{-2} should be changed to the correct symbol, upper case 'M'. That is

460 HL3 = 1.7*0.00001*M*(5867 − PA) which is the formula for the latent respiration heat loss

470 HL4 = 0.0014*M*(34 − TA) which is the formula for dry respiration heat loss

510 TS = 0.303*EXP(−0.036*M) + 0.028 the relationship between thermal sensation and thermal load

In line '480' a bracket should be added after '100' as follows

480 HL5 = 3.96*FCL*(XN^4 − (TRA/100)^4) which is the formula for heat loss by radiation

All of these corrections are made for the computer program described below. However, the reader may wish to use the standardised program used in ISO 7730 (2005) and will require to implement the corrections provided above.

THE INTERFACE

The interface of any computer program will need to take account of user requirements. The use of the program will determine the inputs and outputs required and the form in which they will be received and presented to the user. A possible interface is shown in Figure 10.1. This interface is in spreadsheet format and is easily set up as a shell outside of the program (blue boxes, labels, etc.)

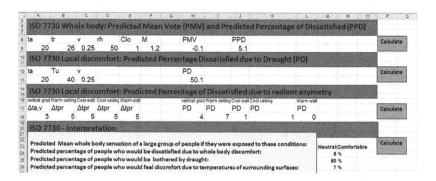

FIGURE 10.1 A computer interface using a spreadsheet to determine the PMV and PPD thermal comfort indices and local thermal discomfort.

and will appear when the spreadsheet is opened. The inputs and outputs will appear in the boxes under the labels as replacements to the default values provided in the shell spreadsheet. This interface is just an example and can be simplified, changed or elaborated upon. It can be created in the computer program or as it is here as a shell outside of the computer program. The essential point is that it starts with a spreadsheet.

This interface covers all of the thermal comfort predictions in ISO 7730 (2005) including the PMV/PPD indices as well as Percentage Dissatisfied predictions from different forms of local thermal discomfort (see Chapters 2, 3 and 5). Actually the local thermal discomfort calculations are simple equations calculated directly as formulae on the spreadsheet. They are not included in the 'macro' code which is exclusively used here for the calculation of the PMV and PPD indices and considered below. For the PMV and PPD calculations the inputs are air temperature (t_a), mean radiant temperature (t_r), air velocity (v), relative humidity (rh), clothing (clo) and metabolic rate (mets). For this interface the inputs can be provided by answering questions in pop-up boxes but a simpler method is to type in the values in the appropriate places on the spreadsheet. The values are then linked to the variables in the computer program by the following instruction:

```
Range("A9").Select: TA = ActiveCell.Value: Application.
ActiveSheet.Range("A9").Value = TA
```

So the value of air temperature is given the symbol TA in the computer code and taken from spreadsheet box A9 (a value of 20 in Figure 10.1). Similar inputs are for TR from B9; VEL from C9; RH from D9; CLO from E9, M from F9 and WME from G9.

Outputs are for PMV in H9 using the code: Application.ActiveSheet. Range("H9").Value = PMV: and similar for PPD in J9. During the development of the program, it is useful to implement similar code to output calculated values as a check that the code is correct. These can be removed for the final interface design.

CALCULATION OF PMV AND PPD

Preliminary Calculations, Unit Conversions and Assumptions

A calculation of the saturated vapour pressure at any temperature is given by Antoine's equation. The partial vapour pressure in the air in KPa is the relative humidity multiplied by 10 to convert mb to KPa multiplied by the saturated vapour pressure at air temperature (TA). The computer code is as follows:

```
PA = RH*10*Exp(16.6536 - 4030.183/(TA + 235))
```

ICL is clothing insulation in m²°C W⁻¹: ICL = 0.155*CLO
Metabolic rate of 1 Met is 58.15 W m⁻²: M = MET*58.15
Work rate of 1 Met is 58.15 W m⁻²: W = WME*58.15
Note that WME is often assumed to be 0.
Metabolic heat production is metabolic rate minus any mechanical
   work: MW = M – W

Estimate of the increase in surface area due to clothing from empirical equations

```
If ICL < 0.078 Then FCL = 1 + 1.29*ICL: Else: FCL = 1.05 +
0.645*ICL: End If
```

The heat transfer coefficient by forced convection is 12.1 multiplied by the square-root of the air velocity: HCF = 12.1*Sqr(VEL)

To convert temperature in °C to temperature in K add 273: TAA = TA + 273: TRA = TR + 273

'CALCULATE SURFACE TEMPERATURE OF CLOTHING BY ITERATION'

The problem is that the surface temperature of clothing (TCL) is calculated from an equation that contains the surface temperature of clothing (TCL). In non-computer symbols

```
tcl = 35.7 - 0.028(M - W) - I_cl × {3.96 × 10^-8 × fcl ×
[(t_cl + 273)^4 - (t_r + 273)^4] + f_cl × hc × (t_cl - t_a)}
```

The solution is to make an educated guess at the surface temperature of clothing (TCLA), put the value into the equation and use any discrepancy (EPS) to modify and improve the guess so that the value converges to a value that satisfies the equation (that is within a value of 0.00015 in this program).

First guess for the surface temperature of clothing

```
TCLA = TAA + (35.5 - TA)/(3.5*ICL + 0.1)
```

P1–P5 are stages in the calculation:

```
P1 = ICL*FCL: P2 = P1*3.96: P3 = P1*100: P4 = P1*TAA:
P5 = 308.7 - 0.028*MW + P2*(TRA/100)^4
XN = TCLA/100: XF = XN: N = 0: EPS = 0.00015
```

N is set to 0 and the iteration starts.

```
350 XF = (XF + XN)/2
```

Calculation of the heat transfer coefficient by convection as the larger of that for natural and forced convection

```
HCN = 2.38*Abs(100*XF - TAA)^0.25: If HCF > HCN Then
HC = HCF Else HC = HCN
```

The equations are arranged such that a value XN converges to XF. If XN = XF the value of TCL can be calculated from TCL = 100*XN − 273.

```
XN = (P5 + P4*HC - P2*XF^4)/(100 + P3*HC)
N = N + 1
```

The number of iterations is set to a maximum of 150.

```
If N > 150 Then GoTo 550
```

If the difference between the two values is not sufficiently small, then we go back to the start of the iteration where an improved 'guess' is made.

```
If Abs(XN - XF) > EPS Then GoTo 350
```

Line 350 is back to the start of the iteration as the difference is too large and the equation has not been solved.

So if XN-XF is within a small value of 0.00015, then we can calculate the mean surface temperature of clothing TCL as

```
TCL = 100*XN - 273
```

We can now calculate the components of heat loss. HL1 is the heat loss through skin diffusion and includes the skin temperature required for comfort.

```
HL1 = 3.05*0.001*(5733 - 6.99*MW - PA)
```

HL2 is evaporative heat loss by thermoregulatory sweating. This will only apply if activity is greater than at rest and will be at the sweat rate required for comfort

```
If MW > 58.15 Then HL2 = 0.42*(MW - 58.15): Else HL2 = 0!
```

HL3 is the evaporative heat loss due to breathing. It is related to metabolic rate M and hence activity and the difference between the saturated vapour pressure at exhaled air temperature and the partial vapour pressure in the air (related to humidity).

```
HL3 = 1.7*0.00001*M*(5867 - PA)
```

HL4 is the convective heat loss due to breathing. It is related to metabolic rate M and hence activity and the difference between the exhaled air temperature (34°C) and the air temperature (TA).

```
HL4 = 0.0014*M*(34 - TA)
```

HL5 is the heat transfer by radiation. It is related increase in surface area due to clothing (FCL) multiplied by the difference in the fourth powers of the absolute temperatures of clothing (XN) and the mean radiant temperature.

```
HL5 = 3.96*FCL*(XN^4 - (TRA/100)^4)
```

HL6 is the heat transfer by convection. It is FCL multiplied by the convective heat transfer coefficient (maximum of natural or forced see above) multiplied by the difference between clothing surface temperature (TCL) and air temperature (TA).

```
HL6 = FCL*HC*(TCL - TA)
```

'CALCULATE PMV AND PPD'

The PMV is derived from the thermal load, which is the metabolic heat production minus the heat transfer between the person in thermal comfort and the environment (MW minus the sum of the HL values). The thermal load is then multiplied by an empirical equation (TS) relating thermal sensation with metabolic rate. PMV = 0 implies comfort and the value is capped at ±3. For PMV > 3 a spurious 999 value is given and 100% dissatisfied. The PPD is calculated from the PMV value.

```
TS = 0.303*Exp(-0.036*M) + 0.028
PMV = TS*(MW - HL1 - HL2 - HL3 - HL4 - HL5 - HL6)
If Abs(PMV) > 3 Then GoTo 550
PPD = 100 - 95*Exp(-0.03353*PMV^4 - 0.2179*PMV^2)
GoTo 570
550 PMV = 999!
PPD = 100
```

The PMV value is placed in cell H9 and the PPD value in J9.

```
570 Range("H9").Select: ActiveCell.Value = PMV
Range("J9").Select: ActiveCell.Value = PPD
```

EXAMPLE OUTPUTS OF THE COMPUTER PROGRAM

It is important when writing a computer program to be confident that it is correct. One way of confirming this is to compare outputs with those provided in ISO 7730 (2005). Some verification can be provided by comparing the results with PMV values provided in Chapter 3, Table 3.1. An example of how the program can be used is provided below.

A Practical Example

Olesen (1985) considers thermal comfort in a theatre designed for a sedentary audience with low air velocity and air temperature = mean radiant temperature. Using $t_a = t_r$; $M = 58.15\,W\,m^{-2}$; $v < 0.1\,m\,s^{-1}$; 50% rh and clo = 1.0 provides a comfort temperature of 23°C. Actually if we assume still air is natural convection then $v = 0.15\,m\,s^{-1}$ and we assume W = 0 then the comfort temperature for PMV = 0 is 23.7°C and for 23°C, PMV = −0.19. For v = 0.1 m s⁻¹ PMV = −0.08 at 23°C. So the outputs depend upon the inputs and 23°C for comfort is as good as any estimate. In the example of Olesen (1985), it is found that during a performance the mean radiant temperature is 4 K higher (27°C) than air temperature. This is due to radiation from one member of the audience to another. To achieve thermal comfort the air temperature is lowered to 21°C. Actually for t_a = 23°C; t_r = 27°C; v = 0.1 m s⁻¹; rh = 50% with 1 clo and 1 Met of activity, PMV = 0.13 for those conditions by calculation using the computer program which again makes no difference.

The example demonstrates that the computer program used in the calculations here and those of Olesen (1985) are probably the same. A full validation would be to compare scenarios as presented in ISO 7730 (2005). It also demonstrates that although there is no problem with using the PMV value to two decimal points in practice the accuracy of the inputs and hence outputs is less than that. In practice that level of accuracy is not needed.

A more beneficial approach to establishing thermal comfort in the theatre is to use the PMV model in an adaptive approach. What are the adaptive opportunities for the people in the theatre audience? Ability to move around is restricted but there may be opportunity to change clothing. Suppose that we assume that clothing can be reduced from 1.0 to 0.6 clo by removing an outer layer. If we assume that the mean radiant temperature is 4 K above air temperature we can provide the range of comfort conditions from t_a = 21°C and t_r = 25°C for 1.0 clo to t_a = 24°C and t_r = 28°C for 0.6 clo. So a range of air temperatures between 21°C and 24°C is possible if we ensure that it is convenient to remove clothing layers. Actually there are other solutions. Suppose we can't remove clothing then we can maintain comfort using an increase in air velocity to 0.25 m s⁻¹ and keep the air temperature at 23°C. This air velocity may be too high for a theatre however. The demonstration is that there are many solutions that can be achieved with adaptive opportunities and that as a starting point the PMV or other rational index will be able to provide a tool for design. It will allow insightful thinking for design so that solutions such as the use of displacement ventilation, local ventilation or ventilated seats, chilled ceiling, etc. can all be tested in computer-aided design using the simple computer program without resorting to more complex CFD (computational fluid dynamics) techniques.

The identification and prediction of the effects of adaptive opportunities is important for practical application. This must include consideration of the characteristics of the people in the theatre audience. People with physical disabilities may not be able to adjust clothing or move away from uncomfortable conditions, for example, and in some cases unacceptable environments may become regarded as cases of discrimination.

The Thermal Comfort Survey

11

ASSESSMENT AND EVALUATION OF EXISTING ENVIRONMENTS

People sometimes complain that the thermal environment is too hot or too cold or because they are uncomfortable due to local thermal discomfort such as that caused by a draught. Faced with a request to assess an environment for human occupancy and make recommendations, a practitioner may need to conduct an environmental survey. This may involve total environments and include noise, air quality, lighting, thermal environments and more. The following, however, will be restricted to a consideration of how to conduct a survey specifically concerned with thermal comfort. It presents the principles behind a survey as well as practical orientation, where, when and what measure, who to involve and how to analyse and interpret results through to recommendations. For more detailed consideration, see Wilson and Sharples (2015), Nicol et al. (2012), EN 15251 (2007) and ISO 28802 (2012). EN 15251 (2007) "Indoor environmental input parameters for design and assessment of energy performance of buildings addressing indoor air quality, thermal environment, lighting and acoustics" provides general and specific advice and is particularly useful when assessing environments for energy use. ISO 28802 (2012) "Ergonomics of the physical environment – Assessment of environments by means of an environmental survey involving physical measurements of the environment and subjective responses of people" presents an environmental survey method that can be used to assess indoor and outdoor environments.

Research into thermal comfort using field studies involving surveys of people in 'real' environments requires careful consideration and experimental design in order to find conditions in which people find thermal comfort. It has the advantage over laboratory studies (where thermal comfort models can be built from 'artificially' controlled environments with confidence of cause and effect in that context) of investigating realistic conditions but the disadvantage of confounding variables (never quite sure of the cause). Both research methods have their role in research into thermal comfort despite frequent misguided arguments of for and against. This is a discussion for another day. Nicol et al. (2012) provide a description of how to conduct thermal comfort field studies. A particularly influential field study into thermal comfort, 'the Watson House survey', was meticulously carried out over a year by Fishman and Pimbert (1978) (see also McIntyre, 1980). An extensive review of thermal comfort surveys conducted across the world is provided by de Dear (1998) and de Dear et al. (2013).

A less academic but important practical requirement is to conduct an environmental survey to identify whether and why people experience discomfort and to identify the reasons and make recommendations for improvement. Actually this is part of continued environmental design and evaluation and would include post-occupant studies in buildings and other environments. There are many common features of the field study and the environmental survey. It is the practical assessment of an existing thermal environment with the purpose of providing an evaluation of thermal comfort and making recommendations for improvement that is considered below.

As an environmental ergonomist I was often asked to conduct environmental surveys mainly for offices where workers were complaining of discomfort. This is the scenario of the case study presented by Parsons (2005a) where workers in a large office were complaining that their thermal environment was unacceptable. This was consultancy work. One day was allowed for the assessment and a total of 4 days for the whole project including measurement, analysis and final report. There are many variations so each environmental survey will be bespoke. However, a presentation of general principles supported by examples for possible application is presented below and will provide a sound basis for the design and application of the environmental survey.

THE THERMAL COMFORT SURVEY

ISO 28802 (2012) presents principles and methods for assessing physical environments concerned with acoustic, visual and lighting, thermal and air quality environmental components related to the comfort and well-being of people in those environments. The sections relevant to thermal comfort are presented here along with other suggestions and recommendations relevant to conducting a thermal comfort survey.

THE INITIAL BRIEF – BE CLEAR ABOUT THE AIM AND GATHER INFORMATION

The first essential step to any survey is to be specific about the aims of the survey and hence define the types of outcome it will produce. The design of the survey will identify when, where and what to measure as well as how it will be measured, who to involve and how the results will be presented, analysed and interpreted to ensure that the objectives will be achieved. The aims should be agreed with the 'client' who in an initial interview should also be asked about the specific problems and context. The social and political context will be important to judge if there is any possible bias in responses and the level of cooperation the survey will receive. Information on heating, air conditioning and ventilation systems and how people are affected and can interact with them will be useful as well as a description of circulation of air from intake of fresh air to exhaust.

MEASUREMENT OF THE THERMAL ENVIRONMENT

Where, When and 'Who' to Measure

A decision must be taken on where, when, who and what to measure. Where to measure the environment and when to measure it are a matter of statistical sampling. This also applies to who to involve in the survey if it is not possible to involve all people who occupy the space. Samples must be representative and avoid bias. The environment will vary throughout a space and continuously with time. Depending upon the work, people will move throughout the space and over time (and personnel will change with shift work, for example). In general, the more places and times, the more valid the survey. As we are interested in thermal comfort, we should measure where the people are and at what times they are or could be. For a homogeneous distribution of people we could use a grid system. Measuring at workplaces is important and for a simple first survey, discrete measurements are often made (although continuous data logging techniques may be used for more detail of time variations). For a simple 1-day survey, measurements should include when and where complaints had been made, morning or afternoon or if possible both. A first request would be to obtain a scale plan of the 'office' and select measuring positions. Ankle, chest and head height measurements should be taken if local thermal discomfort (e.g. from draughts) is suspected. For subjective measures, it is better to ask people at their workstations how they feel 'NOW' rather than rely on memory of how they felt some time ago.

What to Measure

There are three types of measurement that are useful. Measurement of the physical environment is useful for further analysis and comparison with regulations, similar environments elsewhere as well as what is considered comfortable in standards. Subjective measures provide direct measures of how people feel 'NOW' and any general comments that they have about the environment as well as acceptability and preference. A checklist of observations about the environment and occupants can be completed by the experimenter. This will include observations about the space (e.g. windows, orientation of the sun, etc.), clothing, activity and perceived adaptive opportunities. Can people adjust clothing and move around, for example. Behavioural measures are useful but more difficult to quantify and relate cause to effect than subjective measures which are specific to the thermal environment. However, it will be useful to observe occupant behaviour in the office and identify adaptive opportunities and who uses them and how and when they are used. For a more detailed discussion, see ISO 28802 (2012), Parsons (2000) and Wilson and Sharples (2017). Physiological measures are not usually used in thermal comfort surveys unless more extreme environments are suspected. Local skin temperature measurements, such as on the fingers, may indicate vasomotor thermoregulatory control affecting the extremities, in moderate environments, but offer no significant additional insight.

The minimum physical measures of the environment to be taken in a thermal comfort survey are air temperature, radiant temperature, humidity and air velocity. It is also important to estimate at the position of each set of measurement, clothing insulation and metabolic rate due to activity and also note any adaptive opportunities that are available and those that are used.

MEASUREMENT OF THE THERMAL ENVIRONMENT

The specification and use of instruments for measuring the thermal environment are provided in ISO 7726 (1998). Human thermal comfort is determined by the combined interaction of air temperature, radiant temperature, air velocity, humidity, activity level and clothing on a person. To assess thermal comfort, we need to measure or estimate values of all of those six factors. All of the factors are variables which vary in space and continuously change with time, however, they are often defined as single parameters which are representative of the changing environment and are therefore termed the six basic parameters. An additional seventh factor is adaptive opportunity. All need to be considered in a thermal comfort survey.

Air Temperature (t$_a$)

In the context of human thermal environments, air temperature can defined as "the temperature of the air surrounding the body that drives heat transfer between the body and the air". It should not be measured too close to the body as the air temperature will be influenced by body temperature and not too far away as that may not be representative of air that is driving heat transfer and affecting the person.

Air temperature is measured using calibrated thermistors or thermocouples. Mercury in glass thermometers are not now recommended, as breakage will release toxic mercury. Sensors can be affected by both air temperature and radiation. Rapid movement of air across the sensor, such as when whirling a hygrometer, use of fans or in ventilation systems, external on moving vehicles or outdoor wind, minimises the contribution of radiation. Shielding the sensors with silvered open-ended cylinders reduces the contribution of radiation but they must allow a free flow of air across the sensor and must not heat up, causing a re-radiation effect.

ISO 7726 (1998) provides a specification of the required accuracy for instruments for measuring air temperature related to thermal comfort. Air temperature sensors for measuring thermal comfort should have a measuring range from 10°C ± 0.5°C up to 40°C ± 0.5°C guaranteed at least for a deviation of $|t_a - t_r| = 10°C$. Response time should be as short as possible and a mean value over 1 minute is desirable. The sensor should be effectively protected from radiation.

Mean Radiant Temperature (t$_r$)

Mean radiant temperature can be defined for a point or for an object. At any point in a radiant field, it is defined as the temperature of a uniform enclosure in which a small black sphere would have the same radiant exchange as it does with the real environment. It can be measured using a black sphere with a temperature sensor at its centre, called globe temperature and in effect integrates the effects of radiation from all directions (three dimensional) into a single number. The sphere is black to absorb and emit all wavelengths of radiation.

Globe thermometers will achieve steady-state temperature when convective heat loss or gain is equal to (balanced by) radiant heat gain or loss. If there is no net radiation gain or loss, globe temperature will stabilise at air temperature. When there is a radiation gain or loss, it is important to correct for air temperature and air velocity to obtain mean radiant temperature, as they will also influence globe temperature (ISO 7726, 1998; Parsons, 2014). The correction for the effects of air temperature and air velocity will depend upon the diameter of the globe (often 150 mm, but smaller less accurately corrected globes are often used for convenience). The smaller the globe, the greater the influence of air temperature and air velocity on globe temperature. Any inaccuracies in measurement of air temperature and air velocity will therefore be carried into the correction of globe temperature to obtain mean radiant temperature at a point in space. The size of the globe will have more influence on globe temperature the greater the air velocity and deviation between radiant and air temperature. Often for thermal comfort conditions, therefore, any effect of globe size is small.

Globe thermometers are usually made out of copper for a relatively quick response but the material type will affect only the response time (typically over 15 minutes for a 150-mm diameter black globe, even for metal) not the equilibrium value, although the value obtained will be affected by any changes during the measurement time. A smaller diameter globe will achieve steady state more quickly than a larger globe.

The mean radiant temperature for an object, such as the human body, will depend upon its shape (and hence body size and posture) which will influence the areas projected towards radiation surfaces. The integration of projected areas (Ap) over all directions provides the area available for radiant exchange between the surrounding environment and the person (Ar). Estimates of Ar values divided by the total surface area of the body (Ar/A$_D$) are for standing (0.77), sitting (0.72) and crouching (0.67).

ISO 7726 (1998) provides a specification for sensors for measuring mean radiant temperature. For thermal comfort assessment, sensors should measure from 10°C ± 0.2°C up to 40°C ± 0.2°C. This level of accuracy is beyond that of a globe thermometer which will have a long response time and will therefore only give an indication of average radiation effects. More responsive instruments such as pyranometers can provide directional radiation values that can be integrated to provide mean radiant temperature, as well as measurement of solar radiation values. These are often outside of the scope of, and not necessary for, a first survey for thermal comfort.

When comparing direction of radiation, plane radiant temperature (tpr) is often used. That is "the uniform temperature of an enclosure where the radiance on one side of a small plane element is the same as in the non-uniform actual environment". This measures the radiation for the projected area (Ap) over one direction (up/down, left/right, fore/aft). For example, for a standing man the up/down projected area (Ap/Ar) is low (0.08). Radiant asymmetry is related to local thermal discomfort (see Chapter 5). It is the difference between opposite plane radiant temperatures (e.g. front-back). Plane radiant temperature will supplement information provided by mean radiant temperature.

Air Velocity (v_a)

Air velocity is the speed and direction of air moving across a person. Often an estimate of average air speed is taken. Convective and evaporative heat transfer to and from the body are sensitive to air velocity and measurement is required to less than 0.1 m s^{-1}. Specifications for measuring instruments (anemometers) are given in ISO 7726 (1998).

Air velocity sensors for measuring thermal comfort should have a measuring range of 0.05 to $1.0 \pm (0.02 + 0.07 \, v_a) \text{ m s}^{-1}$. The levels should be guaranteed whatever the direction of flow. Response time should be as short as possible to measure variations in velocity. An indication of the mean value over 3 minutes is desirable and, if a time-varying signal is recorded, an estimate of standard deviation, which will allow a calculation of turbulence intensity (ISO 7726, 1998). Visualisation of air movement using smoke or bubbles, with moving photography (e.g. from a mobile phone), will provide insight into the speed and direction of air.

Humidity (ϕ)

Humidity is the concentration of water vapour in the air and is often expressed as partial vapour pressure (P_a). Relative humidity (ϕ, or sometimes rh) is the partial vapour pressure in the air divided by the saturated vapour pressure at air temperature (P_a/P_{sa}). The dew point (tdp) is the temperature at which the air would become saturated, below which water would condense (clouding up cold widows) and it would start to 'rain'.

Partial vapour pressure can be measured directly using calibrated electronic (capacitance) sensors or can be derived from a psychrometric chart using aspirated (sensors in high air velocity) wet bulb (twb) and dry bulb (tdb) temperatures, from a whirling hygrometer, for example. The hygrometer contains two thermometers one with a soaked (continuously from a distilled water reservoir) wet wick covering its sensor and the other with an open dry sensor. By whirling (rotating longitudinally) the thermometers at high speed in a frame with a rotating handle, the effects of radiation become negligible so dry bulb temperature (tdb) becomes air temperature (t_a) and the wet wick evaporates water and cools the sensor to a temperature which is related to the humidity in the air. A psychrometric chart (Ellis et al., 1972) provides partial vapour pressure, relative humidity and dew point (see Figure 6.2). Dew point can also be measured optically by cooling a silvered surface until dew forms (hence deflecting a reflected light beam). ISO 7726 (2001) provides a specification of instruments for measuring humidity in the assessment of thermal comfort.

Absolute humidity is measured as the partial vapour pressure of water vapour. Sensors for measuring thermal comfort should have a measuring range 0.5–3 ± 0.15 kPa and as short a response time as possible (ISO 7726, 2001).

MEASUREMENTS SHALL REPRESENT ENVIRONMENTS TO WHICH PEOPLE ARE EXPOSED

It is essential that the instruments are placed such that they measure the environment that would be experienced by the people in the space (e.g. on a desk) but must not interfere with the task, the environment they are 'attempting to measure', and any subjective or observational measures. The main consideration is that the environment being measured is that experienced by the person. This includes the spatial distribution (three dimensions) as well as temporal distribution. This is a non-trivial point as instruments (and their containers and packaging) can block out the sun, create an artificial context, etc. Experimenter interference (leaning over an instrument, creating a social context and atmosphere, etc.) must be insignificant.

Particular consideration must be given to the characteristics of the instruments and what they are actually measuring. Calibration procedures (before and after to check for drift) must be carried out and reported. The range, accuracy, sensitivity, time constant, reliability and robustness must all be considered and appropriate for the measurements being performed.

Improvisations to supplement measurements are often useful (note outside weather conditions, windows open or closed). I have found that the use of bubbles (children's bubbles are easily accessible) provides great insight into air movement (including potential draughts). They can, however, be 'messy' involving soap containers, drips and bursts.

Clothing and Activity

Clothing insulation can be estimated from ISO 9920 (2007) but for most purposes an estimation between 0.5 clo for T-shirt and shorts to 0.75 clo for long trousers or skirt and long-sleeved shirt to 1.0 clo for a typical business suit including jacket will provide a reasonable first approximation. For the survey, it would be useful to monitor clothing behaviour and itemise garments at the time measurements and subjective responses are taken. Also at that time the activity of the person should be noted and an estimate made from 1 Met for sitting at rest to 1.2 Met for standing at rest and light work to greater values for higher activity. For more dynamic work a description of tasks and movement across the space with time will be important.

Adaptive Opportunity

How people behave to achieve thermal comfort or avoid thermal discomfort is a major consideration and it is essential that it is captured in some way in a thermal survey. Possible adaptive opportunities and if and how they are used (focus groups and interviews are useful for this) as well as potential adaptive opportunities (relax clothing code, allow people to move around, etc.) and how they will affect environmental conditions to which people are exposed are all important. This should be directly addressed in any initial briefing with the client and in recommendations. The use of widely available moving digital photography (even infra-red photography) to investigate adaptive behaviour has great potential but must consider ethical issues.

Subjective Measures

Subjective measures are essential for a thermal survey and if not possible to carry out will greatly restrict the validity of any conclusions. I have found that a single sheet questionnaire completed by the person at their workplace (maybe electronically) provides useful data. Comments about thermal comfort in general and particularly at times when the survey was not being carried out are usually a good method for indicating if the survey will be valid and typical. They may indicate that a further survey is required. The following subjective scales are useful and recommended by ISO 28802 (2012). These are categorised by ISO 10551 (2019) as perceptual (sensation), affective (comfort), preference, acceptance and tolerance.

Additional scales and open-ended questions (any other comments? etc.) can supplement the responses on the scale. I often double up the scales by asking "how do YOU feel NOW" and "How do YOU generally feel at work". The emphasis in capitals avoids a person trying to interpret feelings of the group or others in the group. People often want to discuss the issue with others so it is important not to simply obtain a collective response. Conducting focus groups (after subjective measurements are taken from individuals) will be invaluable if context allows. If local thermal discomfort is an issue and particularly if people report being cold, then a diagram of a person can be provided and scales used for areas of the body as well as for overall comfort. Scales can be in the form of categories or more usefully as a continuous form where people mark the position on a line marked with evenly spaced labelled category marks (warm, hot, etc.). McIntyre (1980) provides a comprehensive review of the properties of subjective scales used for assessing the thermal environment. He notes that the Bedford scale (much too warm, too warm, comfortably warm, comfortable-neither warm nor cold, comfortably cool, too cool, much too cool) confounds comfort and sensation as well as preference. When used in practice, however, it performs similarly to the usually used ASHRAE/PMV/ISO scale of thermal sensation (hot, through neutral to cold – see below). Hodder and Parsons (2001b) found a high correlation between responses to different sematic scales for assessing different aspects of the thermal environment (e.g. sensation and comfort). McIntyre (1980) proposes the additional use of a preference scale (warmer no change, cooler) with a sensation scale.

Subjective scales typically used for assessing thermal comfort (with respect to YOUR thermal environment how do YOU feel NOW) are thermal sensation (hot, warm, slightly warm, neutral, slightly cool, cool, cold); uncomfortable (not uncomfortable, slightly uncomfortable, uncomfortable, very uncomfortable); stickiness (not sticky, slightly sticky, sticky, very sticky); preference (much warmer, warmer, slightly warmer, no change, slightly cooler, cooler, much cooler); acceptability (acceptable, not acceptable); satisfaction (satisfied, not satisfied); draughtiness (not draughty, slightly draughty, draughty, very draughty); dryness (not dry, slightly dry, dry, very dry). From the basic scales a questionnaire can be produced. The questionnaire should be explained to the people completing it. The exact question asked should be clear and leading questions should be avoided (please confirm that you are comfortable, etc.) (see Sinclair, 2005). An example of a simple questionnaire is provided in Parsons (2005b).

Qualitative Methods

As the study and provision of thermal comfort moves to understanding individual and group behaviour, in addition to the physical measurement of the
environment and of human responses, the thermal comfort survey will require
both quantitative (measurement) and qualitative methodologies to provide a
comprehensive understanding of people's reaction to physical environments.
Hignett (2005) provides an overview of qualitative methodologies from a
practical perspective. The use of focus groups, interviews, discourse analysis,
observational techniques and more will provide insight into thermal comfort
responses from an holistic human and individual centred perspective. The
understanding of individual thermal comfort and how it makes up the whole
group response will provide a wealth of information, otherwise untapped, for
environmental design. This approach, however, is in its infancy and how it can
be used to most benefit in a thermal comfort survey has yet to be determined.

PRESENTATION, ANALYSIS AND INTERPRETATION OF DATA

Physical Measurements

The physical measurements at each workplace should be presented on
a plan of the room to provide a first indication of environments people are
experiencing. These can be compared with criteria and objectives for any
heating, air-conditioning and ventilation systems (if they exist). The subjective
measures should also be presented on a plan of the space. Much of the interpretation can be done by visualising the data and reporting particular features.

If integrated with estimates of clothing insulation and metabolic heat production the physical measurements can be used to calculate the values of the
Predicted Mean Vote (PMV) and Predicted Percentage of Dissatisfied (PPD)
indices according to ISO 7730 (2005). Although the PMV is the Predicted
Mean Vote that would be provided by a large group of people had they experienced the conditions measured, an interpretation can be used to give recommendations for each workplace. Fanger (1970) suggested the use of the Lowest
Possible Percentage of Dissatisfied (LPPD) as the best that can be achieved by
gross changes in the environment, for example, by adjusting air temperature.
Plotting the PMV values on a plan of the space will give an indication of
a more detailed set of requirements across the space. Local discomfort will

be indicated from the measurements and questionnaire results and the general questions will indicate if the findings were typical over time. Criteria for thermal environment acceptability is often taken as a PMV value of between slightly warm and slightly cool (PMV = +1 through 0 at neutral to −1). More recent suggestions propose limits of PMV = ±0.7 to ±0.3 for higher quality environments.

The PMV index provides a versatile way of comparing engineering solutions to providing thermal comfort (altering air temperature, air movement, radiation levels and combinations, etc.). It can also provide ranges of air temperature, for example, for thermal comfort if adaptive opportunities are included in the analysis and interpretation. Allowing adjustment to clothing and moving around will indicate the range of conditions that will still allow comfort criteria to be met. In future surveys, it is likely that rather than recommend simply changing environmental parameters to achieve comfort conditions through environmental design and engineering measures, adaptive approaches could be recommended which would have the advantage of user satisfaction as well as energy efficiency.

ISO 28802 (2012) notes that the responses of people that will be measured will depend upon the aims of the environmental survey. Objectives are usually concerned with thermal comfort but could be related to health and productivity. In some cases, skin temperatures and heart rate may be useful measures. In a novel situation, subjective scales may have to be developed from first principles by investigating psychological continua. The subjective scales presented above, however, are those that have been established and are generally appropriate. How the scales are presented to the people is important. A single sheet questionnaire reduces resistance and this can be administered in an electronic form for immediate analysis. 'Subject' training will be needed even if just a brief explanation. An example response at the top of the questionnaire is useful. Leading questions should be avoided and care must be taken if using scales in English, for people where English is not their first language, and in translation.

A final word is that the attitude and disposition of the investigating team must be to collect genuine, valid data and to use it to make recommendations that will not only avoid thermal discomfort and provide comfort, but to promote healthy, energy efficient, pleasant, delightful and productive thermal environments.

References

ASHRAE, 1966, ASHRAE standard 55-1966, thermal environmental conditions for human occupancy, ASHRAE, Atlanta.

ASHRAE, 1974, ASHRAE standard 55-74, thermal environmental conditions for human occupancy, ASHRAE, Atlanta.

ASHRAE, 1998, Field studies of thermal comfort and adaptation, Technical data bulletin vol 14 no 1, ASHRAE Meeting, San Francisco, USA, January 1998.

ASHRAE, 2017, ASHRAE standard 55-2017 "thermal environmental conditions for human occupancy (ANSI/ASHRAE approved)", ASHRAE, Atlanta.

ASHRAE, 2019, Addendum to ASHRAE 55-2017.

Auliciems, A., June 1981, Towards a psycho-physiological model of thermal perception, *International Journal of Biometeorology*, 25(Issue 2), 109–122.

Brager, G. S., Paliaga, G. and de Dear, R. J., 2004, Operable windows, personal control and occupant comfort, *ASHRAE Transactions*, 110(2), 17–35.

Braun, T. L. and Parsons, K. C., 1991, Human thermal responses in crowds, in Lovesey, E. J. (ed.), *Contemporary Ergonomics*, pp. 190–195, London: Taylor & Francis Group.

Bröde, P., Fiala, D., Blazejczyk, K., Holmer, I., Jendritszky, G., Kampman, B., Tinz, B. and Havenith, G., 2012, Deriving the operational procedure for the Universal Thermal Climate Index (UTCI), *International Journal of Biometeorology*, 56(No 3), 481–494.

Brooks, J. and Parsons, K. C., 1999, An ergonomics investigation into human thermal comfort using an automobile seat heated with encapsulated carbonized fabric (ECF), *Ergonomics*, 42(5), 661–673.

Cabanac, M., 1981, Physiological signals for thermal comfort, in Cena, K. and Clark, J. A. (eds.), *Bioengineering, Thermal Physiology and Comfort*, pp. 181–192, Amsterdam: Elsevier.

Cena, K. and Spotila, J. R., 1986, Thermal comfort for the elderly, behavioural strategies and effort of activities, Final report, ASHRAE, RP-460.

Collins, K. J. and Hoinville, E., 1980, Temperature requirements in old age, *Building Services Engineering Research and Technology*, 1(4), 165–172.

Coull, N., 2019, Thermoregulatory responses and ageing: A body mapping approach, *PhD Thesis*, Loughborough University, UK.

Clark, R. P. and Edholm, O. G., 1985, *Man and His Thermal Environment*, London: Edward Arnold.

de Dear, R. J., 1998, A global database of thermal comfort field experiments, in *Field Studies of Thermal Comfort and Adaptation*, Technical data bulletin vol 14 no 1, ASHRAE Meeting, San Francisco, CA, January 1998.

de Dear, R. J. and Brager, G. S., 1998, Developing an adaptive model of thermal comfort and preference, in *Field Studies of Thermal Comfort and Adaptation*, Technical data bulletin vol 14 no 1, ASHRAE Meeting, San Francisco, USA, January 1998.

de Dear, R. J. and Spagnolo, J., 2005, Thermal comfort in outdoor and semi-outdoor environments, in Tochihara, Y. and Ohnaka, T. (eds.), *Environmental Ergonomics*, pp. 269–276, Amsterdam: Elsevier.

de Dear, R., Akimoto, T., Arens, E. A., Brager, G., Candido, C., Cheong, K. W. D., Li, B., Nishihara, N., Sekhar, S. C., Tanabe, S., Toftum, J., Zhang, H. and Zhu, Y. 2013. Progress in thermal comfort research over the last twenty years, *Indoor Air*. doi:10.1111/ina.12046.

DIN 33 403-5, 1994, Climate at workplaces and their environment, Part 5. Ergonomic design of cold workplaces, German Standard, DIN, Berlin.

Du Bois, D. and Du Bois, E. F., 1916, A formula to estimate surface area if height and weight are known, *Archives of Internal Medicine*, 17, 863.

Ellis, F. P., Smith, F. E. and Walters, J. D., 1972, Measurement of environmental warmth in SI units, *British Journal of Industrial Medicine*, 29, 361–377.

EN 15251, 2007, Indoor environmental input parameters for design and assessment of energy performance of buildings addressing indoor air quality, thermal environment, lighting and acoustics, CEN, Brussels.

Factories Act, 1961, Act of Parliament of the United Kingdom, HMSO, London.

Fanger, P. O., 1970, *Thermal Comfort*, Copenhagen: Danish Technical Press.

Fanger, P. O. and Toftum, J., 2002, Extension of the PMV model to non-air conditioned buildings in warm climates, *Energy in Buildings*, 34(6), 533–536.

Fiala, D., Havenith, G., Bröde, P., Kampmann, B. and Jendritzky, G., 2012, UTCI-Fiala multi-node model of human heat transfer and temperature regulation, *International Journal of Biometeorology*, 56(3), 429–441.

Fishman, D. S. and Pimbert, S. L., 1978, Survey of subjective responses to the thermal environment in offices, in *Indoor Climate Effects on Human Comfort, Performance and Health in Residential, Commercial and Light Industry Buildings*, Copenhagen, WHO Conference.

Fung, W. and Parsons, K. C., 1996, Some investigations into the relationship between car seat materials and thermal comfort using human subjects, *Journal of Coated Fabrics*, 26, 47–176.

Gagge, A. P., 1937, A new physiological variable associated with sensible and insensible perspiration, *American Journal of Physiology*, 120, 277–287.

Gagge, A. P., Burton, A. C. and Bazett, H. C., 1941, A practical system of units for the description of the heat exchange of man with his thermal environment, *Science*, 94, 428–430.

Gagge, A. P., Stolwijk, J. A. J. and Nishi, Y., 1971, An effective temperature scale based on a single model of human physiological temperature response, *ASHRAE Transactions*, 77, 247–262.

Gagge, A. P., Nishi, Y. and Gonzalez, R. R., 1973, Standard effective temperature: A single temperature index of temperature sensation and thermal discomfort, in *Proceedings of the CIB Commission W45 (Human Requirements) Symposium*, Building Research Station, 13–15 September 1972. Watford: HMSO.

Gerrett, N., Ouzzahra, Y., Coleby, S., Hobbs, S., Redortier, B., Voelcker, T. and Havenith, G., 2014, Thermal sensitivity to warmth during rest and exercise a sex comparison, *European Journal of Applied Physiology*, 114, 1451–1462. doi:10.1007s00421-014-2875-0.

Gerrett, N., Ouzzahra, Y., Redortier, B., Voelcker, T. and Havenith, G., 2015, Female sensitivity to hot and cold during rest and exercise, *Physiology & Behaviour*, 152, 11–19.

Giorgi, G., Megri, A. C., Donnini, G. and Haghighat, F., 1996, Responses of disabled persons to thermal environments, ASHRAE Research Project 885RPAND Inc., Montreal, Canada.

Griefahn, B., Kunemund, C. and Gehring, U., 2001, Annoyance caused by draught the extension of the draught rating model (ISO 7730), in *Proceedings of the Windsor Conference 2001: Moving Thermal Comfort Standards into the 21st Century*, pp. 135–145, Windsor.

Hardy, R. N., 1979, Temperature and animal life, in *Studies in Biology*, 2nd edn, vol 35, pp. 83, London: Edward Arnold.

Hashiguchi, N., Feng, Y. and Tochihara, Y., 2010, Gender differences in thermal comfort and mental performance at different vertical air temperatures, *Arbeitsphysiologie*, 109(1), 41–48.

Hignett, S., 2005. Chapter 22: Qualitative methodology, in Wilson, J. and Corlett, N. (eds.), *Evaluation of Human Work*, 3rd edn, pp. 113–128, New York: Taylor & Francis Group.

HMSO, 1973, Thermal comfort and moderate heat stress, in *Proceedings of the CIB Commission W45 (Human requirements) Symposium Held at the Building Research Station*, 13–15 September 1972.

Hodder, S. G. and Parsons, K. C., 2001a, *Automotive Glazing, Task 2.4—Thermal Comfort*, Final Technical Report No. BE9733020 T2,4 to BriteEuram Programme, Contract No. BRPRC7970450, European Commission Brussels.

Hodder, S. G. and Parsons, K. C., 2001b, Field trials in Seville, in *Automotive Glazing: Task 2.4—Thermal Comfort*, Final Technical Report No. BE963020, Brite/Euram, European Commission, Brussels.

Hodder, S. G., 2002, Thermal comfort in vehicles the effects of solar radiation, *PhD Thesis*, Loughborough University, UK.

Houghton, F. C. and Yagloglou, C. P., 1923, Determining equal comfort lines, *Journal of ASHVE*, 29, 165–176.

Humphreys, M. A., 1972, Clothing and thermal comfort of secondary school children in summertime, in *Proceedings of the CIB Commission W45 (Human Requirements) Symposium Held at the Building Research Station*, 13–15 September 1972, Watford: HMSO.

Humphreys, M. A., 1978, Outdoor temperatures and comfort indoors, *Building Research and Practice*, 6(2), 92–105.

Humphreys, M. A., Rijal, H. B. and Nicol, J. F., 2010, Examining and developing the adaptive relation between climate and thermal comfort indoors, in *Proceedings of conference on adapting to change. New thinking on comfort*, Cumberland Lodge, Windsor, 9–11 2010. London, Available at http://nceub.org.uk.

Humphreys, M. A., Nicol, F. and Roaf, S., 2014, *Adaptive Thermal Comfort: Foundations and Analysis*, London: Routledge.

ISO 7726, 1998, ED 2, Ergonomics of the thermal environment—Instruments for measuring physical quantities, Geneva: International Organization for Standardization.

ISO 7726, 2001, ED 3, Ergonomics of the thermal environment—Instruments for measuring physical quantities, Geneva: International Organization for Standardization.

ISO 7730, 2005, ED 3, Ergonomics of the thermal environment—Analytical determination and interpretation of thermal comfort using calculation of the PMV and PPD indices and local thermal comfort criteria, Geneva: International Organization for Standardization.

ISO 8996, 2004, ED 2, Ergonomics of the thermal environment—Determination of metabolic rate, Geneva: International Organization for Standardization.

ISO 9920, 2007, ED 2, Estimation of thermal insulation and water vapour resistance of a clothing ensemble (see also Amended Version 2009), Geneva: International Organization for Standardization.

ISO 9920, 2009, ED 2, Estimation of thermal insulation and water vapour resistance of a clothing ensemble (Amended Version), Geneva: International Organization for Standardization.

ISO 10551, 2019, Ergonomics of the thermal environment—Assessment of the influence of the physical environment using subjective judgement scales, Geneva: International Organization for Standardization.

ISO 11079, 2007, ED 1, Ergonomics of the thermal environment—Determination and interpretation of cold stress when using required clothing insulation (IREQ) and local cooling effects, Geneva: International Organization for Standardization.

ISO TS 13732-2, 2001, Method for the assessment of human responses to contact with surfaces (ISO DTR 13732)—Part 2: Human contact with surfaces at moderate temperature, London: BSI.

ISO 14505-2, 2006, Ergonomics of the thermal environment—Evaluation of the thermal environment in vehicles—Part 2: Determination of equivalent temperature, Geneva: International Organization for Standardization.

ISO 14505-3, 2006, Ergonomics of the thermal environment—Evaluation of the thermal environment in vehicles—Part 3: Evaluation of thermal comfort using human subjects, Geneva: International Organization for Standardization.

ISO TR 14415, 2005, Ergonomics of the thermal environment—Application of international standards for people with special requirements, Geneva: International Organization for Standardization.

ISO 16594, 2019, New work project under development by ISO TC 159 SC5 WG1, Designing environments for thermal comfort, Geneva: International Organization for Standardization.

ISO TR 22411, 2019, Ergonomics data and guidelines for the application of ISO/IEC Guide 71 to products and services to address the needs of older persons and persons with disabilities.

ISO NWI 23454-1, 2019 Ergonomics of the physical environment: Human performance in physical environments: Part 1: A performance framework, Geneva: ISO.

ISO DP 23454-2, 2019, Ergonomics of the physical environment: Human performance in physical environments: Part 2: Measures and methods for assessing the effects of the physical environment on human performance, Geneva: ISO.

ISO 28802, 2012, Ergonomics of the physical environment—Assessment of environments by means of an environmental survey involving physical measurements of the environment and subjective responses of people.

ISO 28803, 2012, Ergonomics of the physical environment—Application of international standards to physical environments for people with special requirements, Geneva: International Standards Organization.

Jendritzky, G. and de Dear, R., (eds.), 2012, Special issue, universal thermal climate index (UTCI), *International Journal of Biometeorology*, 56(3), 419.

Jendritzky, G., de Dear, R. and Haventih, G., 2012, UTCI—Why another thermal index? *International Journal of Biometeorology*, 56(3), 421–428.

Jessen, C., 2001, *Temperature Regulation in Humans and Other Mammals*, Berlin: Springer. ISBN 3-540-41234-4.

JIS TR S 0002, 2006, Thermal sensations of young and old persons in moderate thermal environments, Japanese standard.

Karjalainen, S., 2012, Thermal comfort and gender: a literature review, *Indoor Air*, 22(2), 96–109. doi:10.111/j.1600-0668.2011.00747.

Kelly, L. K., 2011, Thermal comfort on rail journeys, *PhD Thesis*, Department of Human Sciences, Loughborough University, UK.

Kennedy, R. S. and Bittner, A. C., 1977, The development of a Navy Performance Evaluation test for Environmental Research (PETER), Presented at Navel Research, San Diego, Available at https://apps.dtic.mil/dtic/tr/fulltext/u2/a056047.pdf.

Kuno, Y., 1956, *Human Perspiration*, Springfield, IL: Charles C Thomas.

Lan, L., Lian, Z., Liu, W. and Liu, Y., 2008, Investigation of gender difference in thermal comfort for Chinese people, *European Journal of Applied Physiology*, 102(4), 471–480.

Lankilde, G., 1977, Thermal comfort for people of high age, *INSERM*, 77, 187–194.

Li, B. and Lim, D., 2013, Chapter 14: Occupant behavior and building performance, in Yao, R. (ed.), *Design and Management of Sustainable Built Environments*, pp. 279–304, London: Springer Verlag.

Liu, Y., Wang, L., Liu, J. and Di, Y., 2013, A study of human skin and surface temperatures in stable and unstable environments, *Journal of Thermal Biology*, 38(7), 440–448.

Loveday, D. L., Parsons, K. C., Taki, A. H., Hodder, S. G. and Jeal, L. D., 1998, Designing for thermal comfort in combined chilled ceiling and displacement ventilation environments, *ASHRAE Transactions*, 104(Part 1B), 901–911.

Madsen, T. L. and Olesen, B. W., 1986, A new method for evaluation of the thermal environment in automotive vehicles, *ASHRAE Transactions*, 92(Part 1B), 38–54.

Marzetta, L. A., 1974, Engineering and construction manual for an instrument to make burn hazard measurements in consumer products, US Department of Commerce, National Bureau of Standards Technical Note 816, Washington, DC.

McIntyre, D. A., 1980, *Indoor Climate*, London: Applied Science.

McNall, P. E., Ryan, P. W., Rohles, F. W., Nevins, R. G. and Springer, W. E., 1968, Metabolic rates at four activity levels and their relationship to thermal comfort, *ASHRAE Transactions*, 74(Part 1. IV), 3.1.

Meese, G. B., Kok, R., Lewis, M. I. and Wyan, D. P., 1984, A laboratory study of the effects of moderate thermal stress on the performance of factory workers, *Ergonomics*, 27(1), 19–43.

Nadel, E. R., Mitchell, J. W. and Stolwijk, J. A. J., 1973, Differential thermal sensitivity in the human skin, *Pflugers Archives*, 340, 71–76.

Nevins, R. G., Michaels, K. B. and Feyerherm, A. M., 1964a, The effects of floor surface temperature on comfort, Part I college-age males, *ASHRAE Transactions*, 17, 29.

Nevins, R. G., Michaels, K. B. and Feyerherm, A. M., 1964b, The effects of floor surface temperature on comfort, Part II college-age females, *ASHRAE Transactions*, 70, 37.

Nevins, R. G., Rohles, F. H., Springer, W. and Feerherm, A. M., 1966, A temperature humidity chart for thermal comfort of seated persons, *ASHRAE Transactions*, 72(1), 283–291.

Nicol, J. F. and Humphreys, M. A., 1972, Thermal comfort as part of a self regulating system, in *Thermal Comfort and Moderate Heat Stress*, in *Proceedings of the CIB Commission W45 (Human Requirements) Symposium, Building Research Station*, 13–15 September 1972. Watford: HMSO.

Nicol, F., Humphreys, M. and Roaf, S., 2012. *Adaptive Thermal Comfort*, Oxford: Routledge, ISBN 978-0-415-69159-8.

Nilsson, H., Holmer, I., Bohm, M. and Noren, O., 1999, Definition and theoretical background of the equivalent temperature. Paper 99A4082, in *Proceedings of the ATA Conference*, Florence.

Nishi, Y. and Gagge, A. P., 1977, Effective temperature scale useful for hypo and hyperbaric environments, *Aviation Space and Environmental Medicine*, 48, 97–107.

O'Brien, N. V., Parsons, K. C. and Lamont, D. R., 1997, Assessment of heat strain on workers in tunnels using compressed air compared to those in free air conditions, in *Mining and Metallurgy, Tunnelling '97 Conference*, pp. 341–352, London: The Institution of Mining and Metallurgy.

Olesen, B. W., 1977, Thermal comfort requirements for floors occupied by people with bare feet, *ASHRAE Transactions*, 83(Part 2), 41–57.

Olesen, B. W., Scholer, M. and Fanger, P. O., 1979, Vertical air temperature differences and comfort, in Fanger, P. O. and Valbjorn, O. (eds.), *Indoor Climate*, pp. 561–579, Copenhagen: Danish Research Institute.

Olesen, B. W., 1982, Thermal comfort, Technical Review, Bruel and Kjaer, Copenhagen.

Olesen, B. W., 1985, Local thermal discomfort, Technical Review No. 1, Bruel and Kjaer, Copenhagen.

O'Neill, D. H., Whyte, R. T. and Stayner, R. M., 1985, Predicting heat stress in complex thermal environments, in Oborne, D. J. (ed.), *Contemporary Ergonomics*, pp. 197–202, London: Taylor & Francis Group.

Oseland, N. A., Humphreys, M. A., Nicol, J. F., Baker, N. V. and Parsons, K. C., 1998, Building design and management for thermal comfort, BRE Report CR 203/98, Garston, UK.

Park, E. S., Park, C., Cho, S., Lee, J. and Kim, E. J., 2002, Assessment of autonomic nervous system with analysis of heart rate variability in children with spastic cerebral palsy, *Yonsei Medical Journal*, 43(No.1), 65–72.

Partridge, R. C. and Maclean, D. L., 1935, Determination of the comfort zone for school children, *Journal of Industrial Hygiene*, 17, 66–71.

Parsons, K. C. and Clark, N. J., 1984, A laboratory investigation of the PMV thermal comfort index, in Megaw, E. D. (ed.), *Contemporary Ergonomics*, pp. 122–127, London: Taylor & Francis Group.

Parsons, K. C. and Bishop, D., 1991, A data base model of human responses to thermal environments, in Lovesey, E. J. (ed.), *Contemporary Ergonomics*, pp. 444–449, London: Taylor & Francis Group.

Parsons, K. C. and Webb, L. H., 1999, Thermal comfort design conditions for indoor environments occupied by people with physical disabilities, Final report to EPSRC research grant GRK71295, Loughborough University, UK.

Parsons, K. C., 2000, An adaptive approach to the assessment of risk for workers wearing protective clothing in hot environments, in Kuklane, K. and Holmer, I. (eds.), *Ergonomics of Protective Clothing*, pp. 34–37, Solna: National Institute for Working Life.

Parsons, K. C., 2002, The effects of gender, acclimation state, the opportunity to adjust clothing and physical disability on requirements for thermal comfort, *Energy and Buildings*, 34(6), 593–600.

Parsons, K. C., 2003, *Human Thermal Environments*, 2nd edn, New York: Taylor & Francis Group.

Parsons, K. C. and Mahudin, N. D. M., 2004, Development of a crowd stress index (CSI) for use in risk assessment, in McCabe, P. (ed.), Contemporary Ergonomics, 410–414, London: Taylor & Francis Group.

Parsons, K. C., 2005a, Chapter 22: The environmental ergonomics survey, in Wilson, J. and Corlett, N. (eds.), *Evaluation of Human Work*, 3rd edn, pp. 631–642, New York: Taylor & Francis Group.

Parsons, K. C., 2005b, Chapter 23: Ergonomics assessment of thermal environments, in Wilson, J. and Corlett, N. (eds.), *Evaluation of Human Work*, 3rd edn, pp. 643–662, New York: Taylor & Francis Group.

Parsons, K. C., Naylor, S. and Wales, J., 2005, The effects of colour and fit of clothing on thermal comfort in the sun, in Bust, P. and McCabe, P. (eds.), *Contemporary Ergonomics*, 2005, pp. 227–231.

Parsons, K. C., 2014, *Human Thermal Environments*, 3rd edn, New York: Taylor & Francis Group. ISBN 978-1-4655-9599-6.

Parsons, K. C., 2018, ISO standards on physical environments for worker performance and productivity, *Industrial Health*, 56(2), 93–95.

Parsons, K. C., 2019, *Human Heat Stress*, Boca Raton, FL: CRC Press. ISBN 978-0-367-00233-6.

Ramsey, J. D. and Kwon, Y. C., 1988, Simplified decision rules for predicting performance loss in the heat, in *Proceedings of a Seminar on Heat Stress Indices*, pp. 337–372. Luxembourg: Commission of the European Communities.

Reegan, A., 2017, Regulating your body temperature with a spinal cord injury, https://ablethrive.com/life-skills/regulating-your-body-temperature-with-a-spinal-cord-injury.

Roaf, S., Nicol, N. and Humpphreys, M. A., 2014, How to design comfortable buildings, Routledge, Oxford.

Rohles, F. H., 1969, Preference for the thermal environment by the elderly, *Human Factors*, 11(1), 37–41.

Rohles, F. H. and Nevins, R. G., 1971, The nature of thermal comfort for sedentary man, *ASHRAE Transactions*, 77(1), 239–246.

Rohles, F. H. and Wallis, P., 1979, Comfort criteria for air conditioned automotive vehicles, Society of Automotive Engineers, SAE INC, 790122.

SAE, 1993, Equivalent temperature. SAE Information report, J2234 1993, Society of Automotive Engineers.

Sinclair, M. A., 2005, Chapter 4: Participative assessment, in Wilson, J. R. and Corlett, E. N. (eds.), *Evaluation of Human Work*, 3rd edn, pp. 83–122, London: Taylor & Francis Group.

Smith, C. J. and Havenith, G., 2011, Body mapping of sweating patterns in male athletes in mild exercise-induced hyperthermia, *European Journal of Applied Physiology*, 111, 1391–1404.

Stennings, P. B., 2007, Thermal comfort of railway passengers, *MPhil Thesis*, Department of Human Sciences, Loughborough University, UK.

Stevens, J. C., Marks, L. E. and Simonson, D. C., 1974, Regional sensitivity and summation in the warm sense, *Physiology and Behavior*, 13, 825–836.

Tanabe, S., Kimura, K. and Hara, T. L., 1987, Thermal comfort requirements during the summer season in Japan, *ASHRAE Transactions*, 93(1), 564–577.

Tochihara, Y., Lee, J.-Y., Wakabayashi, H., Wijayanto, T., Bakri, I. and Parsons, K. C., 2012, The use of language to express thermal sensation suggests heat acclimatization by Indonesian people, *International Journal of Biometeorology*, 56(6), 1055–1064.

Underwood, P. and Parsons, K. C., 2005, Discomfort caused by sitting next to a cold window. Simulated railway carriage at night, in Oborne, D. J. (ed.), *Contemporary Ergonomics*, pp. 23–238, London: Taylor & Francis Group.

Underwood, P., 2006, A practical model for the assessment of thermal comfort in train carriages, *MPhil Thesis*, Department of Human Sciences, Loughborough University, UK.

Van Hoof, J., Bennetts, H., Hansen, A., Kazak, K. J. and Soebarto, V., 2019, The living environment and thermal behaviours of older South Australians: A multi-focus group study, *International Journal of Environmental Research and Public Health*, 16, 935.

Vaughan, S., Martensen, J. and Parsons, K. C., 2005, The effects of intensity of side-on simulated solar radiation on the thermal comfort of seated subjects, in Oborne, D. J. (ed.), *Contemporary Ergonomics*, pp. 435–439, Boca Raton, FL: CRC Press.

Vernon, H. M. and Warner, C. G., 1932, The influence of the humidity of the air on capacity for work at high temperatures, *Journal of Hygiene (Cambridge)*, 32, 431–462.

Webb, L. H. and Parsons, K. C., 1997, Thermal comfort requirements for people with physical disabilities, in *Proceedings of the BEPAC and EPSRC Mini Conference: Sustainable Buildings*, pp. 114–121, Oxford: Abingdon.

Webb, L. H., Parsons, K. C. and Hodder, S. G., 1999, Thermal comfort requirements, a study of people with multiple sclerosis, *ASHRAE Transactions*, 105, 648.

Wong, L. T., Fong, K. N. K., Mui, K. W., Wong, W. W. Y. and Lee, L. W., 2009, A field survey of the expected desirable thermal environment for older people, *Indoor and Built Environment*, 18, 336.

Wyon, D. P. and Holmberg, I., 1972, Systematic observation of classroom behaviuor during moderate heat stress, in *Thermal Comfort and Moderate Heat Stress. Proceedings of CIB, W45 Symposium*, Watford: HMSO.

Wyon, D. P., 2001, Chapter 16: Thermal effects on performance, in Spengler, J. D., McCarthy, J. F. and Samet, J. M. (eds.), *Indoor Air Quality Handbook*, pp. 16.1–16.14, New York: McGraw-Hill.

Wilson, J. and Sharples, S., 2017, *Evaluation of Human Work*, 3rd edn, London: Taylor & Francis Group.

Yao, R., Li, B. and Lui, J., 2009, A theoretical adaptive model of thermal comfort: Adaptive predicted mean vote (aPMV), *Built Environment*, 44, 2089–2096.

Yaglou, C. P. and Messer, A., 1941, The importance of clothing in air conditioning, *Journal of American Medical Association*, 117, 1261–1262.

Yasuoka, A., Kubo, H., Tsuzuki, K. and Isoda, N., 2015, Gender differences in thermal comfort and responses to skin cooling by air conditioners in the Japanese Summer, *Journal of Human-Environment System*, 18(No. 1), 1–20.

Yoshida, J. A., Nomura, M., Mikami, K. and Hachisu, H., 2000, Thermal comfort of severely handicapped children in nursery schools in Japan, in *Proceedings of IEA2000/HFES 2000 Congress*, pp. 712–715, San Diego: International Ergonomics Association.

Index

Printed in the United States
by Baker & Taylor Publisher Services

Printed in the United States
by Baker & Taylor Publisher Services